3D Printing

PRACTICAL GUIDES FOR LIBRARIANS

About the Series

This innovative series written and edited for librarians by librarians provides authoritative, practical information and guidance on a wide spectrum of library processes and operations.

Books in the series are focused, describing practical and innovative solutions to a problem facing today's librarian and delivering step-by-step guidance for planning, creating, implementing, managing, and evaluating a wide range of services and programs.

The books are aimed at beginning and intermediate librarians needing basic instruction/guidance in a specific subject and at experienced librarians who need to gain knowledge in a new area or guidance in implementing a new program/service.

About the Series Editor

The **Practical Guides for Librarians** series was conceived by and is edited by M. Sandra Wood, MLS, MBA, AHIP, FMLA, Librarian Emerita, Penn State University Libraries.

M. Sandra Wood was a librarian at the George T. Harrell Library, the Milton S. Hershey Medical Center, College of Medicine, Pennsylvania State University, Hershey, PA, for over thirty-five years, specializing in reference, educational, and database services. Ms. Wood worked for several years as a development editor for Neal-Schuman Publishers.

Ms. Wood received an MLS from Indiana University and an MBA from the University of Maryland. She is a fellow of the Medical Library Association and served as a member of MLA's Board of Directors from 1991 to 1995. Ms. Wood is founding and current editor of *Medical Reference Services Quarterly*, now in its thirty-fifth volume. She also was founding editor of the *Journal of Consumer Health on the Internet* and the *Journal of Electronic Resources in Medical Libraries* and served as editor/coeditor of both journals through 2011.

Titles in the Series

1. *How to Teach: A Practical Guide for Librarians* by Beverley E. Crane.
2. *Implementing an Inclusive Staffing Model for Today's Reference Services* by Julia K. Nims, Paula Storm, and Robert Stevens.
3. *Managing Digital Audiovisual Resources: A Practical Guide for Librarians* by Matthew C. Mariner.
4. *Outsourcing Technology: A Practical Guide for Librarians* by Robin Hastings

3D Printing
A Practical Guide
for Librarians

Sara Russell Gonzalez
Denise Beaubien Bennett

PRACTICAL GUIDES FOR LIBRARIANS, NO. 22

ROWMAN & LITTLEFIELD
Lanham • Boulder • New York • London

Published by Rowman & Littlefield
A wholly owned subsidiary of The Rowman & Littlefield Publishing Group, Inc.
4501 Forbes Boulevard, Suite 200, Lanham, Maryland 20706
www.rowman.com

Unit A, Whitacre Mews, 26-34 Stannary Street, London SE11 4AB

British Library Cataloguing in Publication Information Available

Library of Congress Cataloging-in-Publication Data
Names: Russell Gonzalez, Sara Ann, author. | Bennett, Denise Beaubien, author.
Title: 3D printing : a practical guide for librarians / Sara Russell Gonzalez, Denise Beaubien
Bennett.
Other titles: Three-dimensional printing
Description: Lanham : Rowman & Littlefield, [2016] | Series: Practical guides for librarians ;
no. 22 | Includes bibliographical references and index.
Identifiers: LCCN 2015041648 (print) | LCCN 2016003070 (ebook) | ISBN 9781442255470
(cloth : alk. paper) | ISBN 9781442255487 (pbk. : alk. paper) | ISBN 9781442255494
(electronic)
Subjects: LCSH: Three-dimensional printing services in libraries. | Three-dimensional printing.
Classification: LCC Z711.96 .R87 2016 (print) | LCC Z711.96 (ebook) | DDC 621.9/88–dc23
LC record available at http://lccn.loc.gov/2015041648

♾™ The paper used in this publication meets the minimum requirements of American
National Standard for Information Sciences—Permanence of Paper for Printed Library
Materials, ANSI/NISO Z39.48-1992.

Printed in the United States of America

Contents

List of Figures and Tables

⊚ Figures

Tables

Preface

We received our first 3D printer in December 2013, played with it while we waited for our charging mechanism, and launched it for patrons in April 2014. In that whirlwind four months, we figured out how to print, trained staff, identified beta testers, set up a discussion list to consult with fellow librarians, and held our breath a lot. We processed 1,200 print jobs during the first year of service, but that represents only a fraction of the related activity. We dismantled the printers to clear jams, tighten belts, and replace parts. We suffered injuries to our fingers until we learned to use appropriate tools instead. We hauled a printer, on a book truck or in a little red wagon, to demonstrations. We created and presented workshops to groups ranging from middle school students to a retirement community to librarians. We coauthored two articles and presented a poster. After countless hours of printer management, staff training, and patron interactions, we are still as fascinated as the newbies to watch a 3D printer in action. Among other comments, when an engineering student asked, "When did 'librarian' become the cool profession?" we knew we made the right call in establishing a 3D printing service.

3D Printing: A Practical Guide for Librarians focuses primarily on what librarians need to know when setting up a 3D printing service in their libraries, acting as a filter for the overabundance of industry news with which librarians are inundated. It includes many tips that we wish we'd known in advance, and it reflects the questions that we have been asked in workshops and presentations. Each chapter includes examples from a wide range of libraries, so you don't have to reinvent the service wheel. Please read the examples from all types of libraries since characteristics of service, such as charging versus not charging or staff-mediated versus patron-initiated printing, are not correlated with library type. We focus almost exclusively on fused filament fabrication (FFF) printers, since those are the type most commonly found in libraries.

By the end of the book, readers should be familiar with the following:

- Rationales for developing a 3D printing service in libraries
- An understanding that 3D printing itself is the end of a process that includes model selection or design
- The basic features and functionality of 3D printers
- The costs and benefits of choosing to offer patron-initiated or staff-mediated printing
- The expense items involved in launching and maintaining a 3D printing service

- Best practices for training, service, and workflow
- How to use the librarians' network for advice and for keeping up with the industry

3D Printing: A Practical Guide for Librarians is composed of twelve chapters that lead readers from an understanding of the technology to best practices for developing a service. Chapter 1, "3D Printing and Its Applications in Libraries," introduces the potential role of a 3D printing service in libraries and provides rationales for launching a service. It includes examples of useful items, not just trinkets, that have been produced by 3D printers. Chapter 2, "3D Printing and Printers," describes the technology of 3D printing. It identifies the printers most commonly used in libraries (as of this writing), provides tips on selecting a printer based on its features and functionality, and outlines the location and space planning needs that a library must consider.

Chapter 3, "Before You Can Print: Getting a Model," describes the three techniques for obtaining a model to 3D print. You can find a model in a repository, use a 3D scanner to reproduce an image of an existing object, or design your own model by using modeling software. In chapter 4, "What Makes a Good Printable Model?" you will learn how to identify whether a model is likely to print successfully before you crank up the printer.

Each library has unique funding sources, user needs, staffing patterns, and location options. Since a 3D printing service involves all of these variables, precise budget scenarios cannot be created. Chapter 5, "Estimating Expenses and Assessing Your Service," presents worksheets with the elements that you need to consider when calculating your budget. As with most library services, "space" plays a huge role in dictating what you can do or afford or where you can put something. The chapter concludes with tips for assessing your service.

Chapter 6, "Developing Your 3D Printing Services," describes several service models that libraries currently offer. It also outlines some additional related services that you may consider offering to your patrons. Chapter 7, "Policy Development," covers two uses of the term "policy." Each library must develop local policies for their service while minding the broader public policy issues of 3D printing, which are being discussed by the American Library Association and which will be helpful to all of us.

Best practices that are unique to library services are covered in chapter 8, "Workflow," and chapter 9, "Preparation and Staff Training." You will learn the specific tasks involved in a 3D printing service, and tips for establishing your staff assignments and staff training program.

Chapter 10, "Dealing with Difficulties," prepares you for coping with the bad stuff that is guaranteed to happen while printer technology is still in its infancy. In chapter 11, "Outreach and Marketing," you will see examples ranging from publicity to workshop outlines that other libraries have used successfully. Chapter 12, "Looking Ahead," summarizes the core planning steps and highlights some short-term trends. Finally, a glossary defines the 3D printing–related terms used throughout this book.

This book doesn't cover everything you need to know about 3D printing, but it should leave you feeling confident in knowing the basic technology, the components to consider in drafting your budget, and the library service issues to resolve. Most importantly, it should remind you that many other libraries are offering a 3D printing service, and you have many colleagues to consult. If you are unfamiliar with the vocabulary of the 3D printing world, please read the glossary before launching into the chapters.

Acknowledgments

We dedicate this book to our long-suffering husbands, Anthony Gonzalez and Rich Bennett, who provided valuable feedback throughout its production. They have also supported our first year of service by listening to our problems and suggesting solutions, serving as assistants during our workshops, and managing our households while we made night, weekend, and vacation pilgrimages to keep our 3D printers humming.

We appreciate the assistance of several coworkers at the University of Florida's George A. Smathers Libraries. In particular, we could not have created a successful 3D printing service that is worthy of publicizing without the buy-in from all of our colleagues at the Marston Science Library. Each of them accepted the new service during a hectic time and quickly or gradually embraced their roles. Hannah Norton and Rachael Elrod established 3D printing services at the Health Science Center Library and the Education Library, and they served to test our assumptions of unique versus universal workflow. Al-ani Ilori assisted with graphic designs included in this book, and Joe Baca and Matthew Daley took the photos in figure 3.2.

Beyond the libraries at the University of Florida, we are grateful to John Loeffler, who initiated a 3D printing service for the College of Engineering and created the model of infill percentages shown in figure 4.6. Our collaboration has resulted in stronger service and the development of a new payment and job management system at both locations.

Several of our favorite patrons inspired us and provided case studies. They submitted models from which we all could learn together. We appreciate the patience and persistence of "Beta Mike," whose quadcopter parts served as our initial learning experience and provided us with the confidence to launch the service. While accepting and processing print jobs, we heard many entertaining backstories about the motivations to create or customize items. 3D printing gives us an opportunity to interact on a more personal level with our patrons. We are humbled when we learn how our patrons' prints make a positive difference in someone's life. We extend our thanks to "Beta Mike," Angelos Barmpoutis, Donald Bejleri, Jessica Bergau, Danny Gonzalez, Claudia Grant, Dylan Wang, and many others. Although we all want to scatter when our Problem Patron approaches, we appreciate his contributions to our internal procedures, many of which now begin with "Always ask . . ." or "Always check . . ."

We are indebted to the creators of library guides and websites that describe their 3D printing services. Many of their details about rationales, funding sources, pricing, service models, policies, workflow, and outreach methods are quoted throughout this book.

We thank the team at the Texas Library Association, especially Edward Kownslar and Elise Walker, for inviting us to present workshops. Developing the outlines of the workshops and this book at the same time strengthened each.

We created the discussion list, Librarymakerspace-L, so we could identify a pool of colleagues to consult on issues surrounding the running of a 3D printing service in libraries. The list currently has over five hundred subscribers. Contributions to this list have found their way into this book in several forms, and our queries to the list for examples to include in the book were met with enthusiasm. In particular, we gratefully acknowledge the contributions of Cody Behles, Deanna Brown, Chris Cooper, Lisa J. Dempster, Jennifer Harris, Amy Jiang, Ashley Kolovitz, Barbara Kountouzi, Liberty McCoy, Adam Rogers, Susan M. Ryan, and Amber Sherman. We extend our apologies to anyone we failed to include here.

Finally, we thank our editor, M. Sandra Wood. She kept us focused on writing for an audience that might have no initial familiarity with the wacky world of 3D printing.

CHAPTER 1

3D Printing and Its Applications in Libraries

> **IN THIS CHAPTER**
>
> ▷ The role of a 3D printing service in a library
>
> ▷ Useful applications of 3D printing
>
> ▷ First steps in planning a service
>
> ▷ Strategies used by libraries to fund their 3D printers
>
> ▷ Options for 3D printing if you're not yet ready to buy a 3D printer
>
> ▷ Methods for keeping up

Why the Library?

3D PRINTING IS AN INCREASINGLY UBIQUITOUS TECHNOLOGY in the media, for education, and for hobbyists. The technology has been available for three decades but, as is the case with many new technologies, the availability in the market of affordable 3D printers (for less than $5,000) suddenly changed the game from fantasy to reality. While 3D printers may not seem affordable or justifiable yet to the average consumer, having access through a collective purchase such as a library or another community-based service is no longer out of reach.

For some observers, a library is not viewed as a logical place for supplying a 3D printing service because it is seen as a departure from the library's mission. A 3D printing service might divert valuable resources, including funding, staffing, and space, from providing information to patrons. However, many supporting arguments have been used to successfully demonstrate that providing access to new technology and new methods of delivering information is actually core to the mission of the library.

A library is a good fit for supporting a 3D printing service, alone or in conjunction with other makerspace activities, because of its support for creative thinking through planned discovery, serendipity, and environments conducive to reflection. In addition to supplying quiet spaces, libraries also encourage collaboration through reading programs and workshops, and by hosting community events. Offering locations and training for DIY makerspaces is a natural extension of these "traditional" activities. Libraries in general are evolving from their traditional role as a knowledge repository to an environment for actively encouraging creation of new knowledge and ideas. Libraries are open to everyone within the user community, and not restricted to "members only" spaces such as specific departments on campus or club members. Some services are offered to guests as well as to primary users. Libraries are known for providing universal access to technology, most obviously with computers and the Internet (Griffey, 2012) but traditionally with photocopiers, printers, and other borrowable devices. And libraries that are struggling in the era of reduced print book circulation to generate ways to keep relevant and to draw in more patrons might find that a 3D printing service will attract an entirely new set of users who then explore other services offered by the library.

Additional arguments include that 3D printing is just another way to create and to visualize information. Finally, a 3D printing service makes a library look engaging and cutting edge, and it provides an opportunity to draw in new patrons to use not only 3D printing but to discover or rediscover additional services. Unlike some classic services about which library staff are the experts, a 3D printing service is exciting because it provides an opportunity for staff and patrons to learn together.

Each library must resolve whether it should offer 3D printing and related services. Perhaps the best starting point is to examine whether a 3D printing service aligns with your library's mission, and chapter 7 reviews how several libraries have integrated the service into their missions by establishing parallel policies. You may also receive strong encouragement or a mandate from administrators, patrons, or potential funders. If so, you may need to gather favorable arguments to convince your staff rather than your funders!

Some libraries place their 3D printers and services in specialized areas, such as in the children's or teens' areas in public libraries (see http://www.richlandlibrary.com/check-it-out/teens) or in the engineering library in academic settings. An appeal to a specialized or limited audience may make sense when a new service is launched.

Sample Rationales

Many libraries post their rationale for providing a 3D printing service on their websites, which may prove helpful for those who could need to gather arguments. Following are some examples to consider when deciding whether your library wishes to introduce 3D printing:

Public Libraries

"Libraries evolve to meet the needs of their communities. Beaufort County Library is evolving into a hub for learning experiences. With the recent rise of STEAM-based education (that's Science, Technology, Engineering, Arts, and Math) we want to provide a place where our community can build those skills through hands-on activities" (Beaufort [SC] County Library, 2015).

"Ultimately, what we are supporting is genuine critical thinking—the move from blind consumption of ideas and things, to the active production and modification of ideas and things, done in community" (Innisfil [Ontario] Public Library, 2014).

In addition to the following statement (Keene [NH] Public Library, 2015), the Keene Public Library's website includes a video (Britton-Smedley, Backus, and Gokey, 2011) that describes the disruptive impact of 3D printing.

The 3D Printer is a tool for innovation and creativity which also happens to be the core of the library. Libraries are places for idea creation and knowledge generation. And here are some ways 3D Printers provide benefits to learning:

- Using math, hard science, engineering and technology to make design prototypes and troubleshoot printer.
- Allows people to be creative with problem solving techniques.
- People have to focus on one project at a time from the beginning to the end.
- And much more.

"Tween/teens identify the library as a place to develop STEAM skills individually or in groups, including use of the Makerspace" (Makerspaces in Idaho Libraries Project, 2015).

"The Library desires to offer community access to new and emerging technologies such as 3D printers to inspire a new interest in design and help the community to bring their creations to life" (Sacramento Public Library, 2013).

Academic Libraries

The Shapiro Library supports Southern New Hampshire University students, faculty, and staff in exploring new technologies, learning new skills, and developing innovation through the equipment, software, and tools available in the Innovation Lab & Makerspace.

Our goals are to foster creativity, innovation, and exploration in a safe and accessible learning environment. It is our sincere hope that members of the SNHU community will use the Shapiro Library Innovation Lab & Makerspace to become:

- Users who can appreciate, understand, and apply various technologies to their everyday life, career and community projects.
- Innovators who can think of new ways to use technologies, to adapt and change them for the betterment of society and their own needs, and to experiment without fear of failure.
- Makers who can build or create new things, concepts, and theories, and collaborate with their peers and colleagues to share experiences, ideas, and support. (Harris and Cooper, 2015b)

"We endeavor to provide the latest technology and tools to enhance research and experiential learning. 3D printers are already starting to be used in departments and programs on campus (such as architecture, engineering, entrepreneurship, information science), but many students and disciplines do not have access to this technology" (University of Arizona Libraries, 2015).

🌀 Useful Applications of 3D Printed Objects

Many people think that 3D printers serve only to generate trinkets, toys, and generally useless objects. Administrators and other potential funders may need to be reassured that

3D printers can be used for creative and productive functions. Below are descriptions of some of the useful objects that have been 3D printed in libraries.

Practical Items

Many patrons use 3D printing to produce replacement parts or tools to assist other functions. The Thingiverse repository of 3D models (MakerBot Industries LLC, 2015) includes categories for "household" and "tools" as well as more whimsical options, illustrating the range of practical objects that could be designed or printed by library patrons. At the Cleveland (OH) Public Library's TechCentral MakerSpace, a patron "designed and 3D printed a unique piece of equipment for his photography business that he was unable to find commercially" (Urban Libraries Council, 2015).

Visualization and Learning

At the University of La Verne, patrons have used the 3D printers to assist students in visualizing difficult concepts and replicating fragile or remote pieces.

> 3D printing helped our faculty visualize some very vague or unique concepts that are hard to explain and demonstrate. For example, we used a 3D printed Menger Sponge to help our math faculty demonstrate the mathematical model of fractal curve [figure 1.1.a]. 3D printing also can help our biology faculty by printing out RNA and DNA objects. These objects go from flat concepts in text or online to concepts in 3D that can now be understood through touch and sight.
>
> A member of the Architecture department approached us, requesting that we print out models of 3D architecture (for example, the Mayan temple at Tikal in Mexico [figure 1.1.b], Parthenon in Athens, Dome of the Rock in Jerusalem [figure 1.1.c], Stonehenge in England [figure 1.1.d]). Previously, they were only able to show students pictures or web slides. This replicated more of an experience they might have out in the field or at a professional firm. Anthropology faculty also asked us to print out archaeology objects for similar reasons; they were able to bring field experiences to the classroom [figure 1.1.e]. For the first time within a classroom setting they were able to have the students handle a 3D model at a low-cost. (Jiang and McCoy, 2015)

Game Design

Southern New Hampshire University has a game design program which teaches students to create digital models of objects and characters. Some Game Design faculty encouraged students to create physical representations of their digital models using the Library's 3D printers. For example, one faculty member explained to students that printed models could be a way to draw attention to their work, especially when applying for employment in the game design field. Additionally, many students wanted a physical representation of their classwork to show family and friends. Towards the end of the academic year, students in some classes were offered extra credit if they brought faculty members 3D printed models of their final project submissions. Library staff learned that translating digital models from the formats required in game design to those that can be 3D printed often required additional considerations. Models used in games and other digital environments do not always require the same construction as models that are built for 3D printing. For example, digital models created in game design classes can involve large files or files with non-manifold geometry which our 3D printing software could not slice.

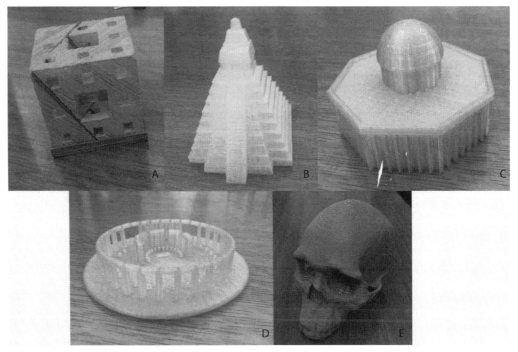

Figure 1.1. 3D Printed Objects from the University of La Verne. *Jennifer Harris*

Library staff worked with students to address potential issues with 3D printing game design models and learned more about the composition of digital models and software which could help correct model issues. (Harris and Cooper, 2015a)

STEAM Education for K–12 Students

The education specialists and scientists at the Florida Museum of Natural History and Duke University are using 3D printing to teach evolution, climate change, and extinction to K–12 students by analyzing fossils such as the teeth of a Megalodon shark, the spine of a Titanoboa snake, and sets of equine teeth. These specimen fossils are fragile and rare and thus not suitable for handling, especially by children. However, by scanning and 3D printing the fossils to create an exact replica, sets are being printed inexpensively and distributed, along with lesson plans, at schools across the country. Additionally, the 3D models are available for download to instructors who have access to a 3D printer at their school or nearby library. For more information, see the project website at http://www.paleoteach.org.

First Steps in Planning a 3D Printing Service

Identify Local Advocates

As you explore the rationale for creating a new 3D printing service, assess its potential users. Supporters who will ensure or promote use may include hobbyists with a backlog of desired jobs, teachers who will craft assignments that require use of the 3D printer, and local makerspaces who are interested in building a community of do-it-yourselfers/DIYers. Those among your supporters who typically show interest in new technologies are your best targets. The University of Florida's proposal (Russell Gonzalez et al., 2013) included brief quotes from faculty in several disciplines who discussed the potential impact of the printers on their teaching and research activities.

Identify Existing Services

Are any 3D printer labs available in or near your community? If so, contact them and ask the following questions:

- What kind of printers and filament do you have?
- Who are your users, and is use restricted to membership?
- What sort of printers/filament/workshops/services do people request that you do not currently supply? And which of these do you think would be viable in this community?

Some existing service providers may sense a hostile takeover or may view the library as competition. Others may be grateful to think that another agency might provide additional services or simply assist with handling the demand. Those who are positive may develop into your strongest allies and become an excellent source of support. Any knowledge or awareness of users' needs and options in your area will guide your planning and strengthen your proposal.

Next Steps

Incorporate the elements noted above if you need to develop and submit a proposal to obtain funding for a 3D printing service. Chapter 2 will introduce you to the technology and to the world of printer options to inform your purchase. Chapter 5 will discuss the array of accessories and space planning that must be considered if you are developing a budget to outfit a new service. If you are fortunate enough to be given a 3D printer, you may manage to bypass the proposal stage and proceed to the other aspects of planning the service.

How Libraries Have Funded Their 3D Printers

Libraries have used a variety of methods to fund the purchase of 3D printers. Following is a sampling of funding sources.

Academic Libraries

"I have found that getting donor funding for 3D printing is some of the easiest fundraising I've ever done. It's an easy sell once a potential donor sees the possibilities. We have made a strong push from the beginning of offering 3D printing to have curricular collaborations with teaching faculty so that we can show the University administration and donors that this is a true learning technology, not just a fun toy. Student and faculty enthusiasm has been high and we have just won our second Library Innovation Award for using the technology collaboratively with faculty for teaching and research" (Ryan, 2015).

"Thanks to a generous sponsorship from the Undergraduate Student Government in May of 2013, . . . after realizing the increased demand and popularity of the service,

University Libraries purchased a second 3D printer and has continued this free service" (Kent State University Libraries, 2015).

The University of Florida submitted a proposal for funding from the university's Student Technology Fee in 2013 and put the first two printers out in 2014. Due to a rapidly changing market, the purchases were considerably different from those drafted in the proposal six months earlier (Russell Gonzalez et al., 2013).

Public Libraries

"We received a state construction grant of $250,000 towards the construction of our FFL Fab Lab space. We also received an Innovation Award of $10,000 from the Contact Summit in October 2011 that was used towards the purchase of equipment, and $13,670 from an IndieGoGo crowdfunding campaign. Our annual operating budget is $1.6 million, and we have strategically reallocated funds away from underutilized resources such as databases and paid performances and lecturers towards the support of hands-on STEAM and making initiatives" (Fayetteville [NY] Free Library, 2014).

"Our Maker Lab was originally funded by an LSTA grant in 2013 when we opened our Central Library. Since then, thanks to generous donations from our community, we've been able to slowly expand the lab to add additional machines" (San Diego Public Library, 2015).

"On January 8th, 2014, the Atlantic City Free Public Library was selected for a $7,500 contract award by the New Jersey State Library and LibraryLinkNJ to kick off an exciting new phase of library service for teens and children. Additional funding will be provided by the Atlantic City Free Public Library, a service of the City of Atlantic City. The initiative, dubbed 'Make AC,' is part of the new wave of Makerspaces hitting libraries all over the nation" (Atlantic City Free Public Library, 2015).

ⓖ Options If You Are Not Ready to Buy a 3D Printer

If your patrons clamor for 3D printing but you do not have immediate plans to obtain your own printer, you have some options besides just saying "no" or "not yet." One option is to consider investing in more basic 3D technology such as the 3Doodler (http://3doodler.com), which allows users to draw with a pen that extrudes plastic filament instead of ink. This type of handheld tool can convey the basic concept of how 3D printing works, and can be used to generate interest in and support for the "real" thing. Staff, patrons, and potential funders need only imagine that a machine, rather than a hand, is guiding the tool in a more precise manner.

You can also refer your patrons to services that will 3D print their files, and you may be surprised to learn of printing services that are available in your area. Table 1.1 lists examples of services that enable potential customers to search for nearby (or distant) printers and to upload their models for printing. Note that the fee is likely greater than what is charged by libraries. Shapeways is a well-established service that will print 3D models, assist with the design process if desired, and also provide a marketplace to print designers' models for a global audience.

Even if you offer a 3D printing service, you may encounter patrons who request that their job be completed earlier than the next spot in your queue, or who will request a type of filament/material or printer functionality that you do not offer. In those cases, a referral to a commercial service may satisfy both patrons and staff.

Table 1.1. Commercial 3D Printing Services

COMMERCIAL 3D PRINTING SERVICE	WEBSITE
3D Hubs	https://www.3dhubs.com/
Makexyz	http://www.makexyz.com/
iMaterialise	http://i.materialise.com/
Sculpteo	http://www.sculpteo.com/en/
Shapeways	http://www.shapeways.com/
updated list of vendors	http://3dprintingforbeginners.com/3d-printing-stores

◉ Keeping Up

General Industry Awareness

Keeping up with any new industry is an overwhelming chore. The market is volatile, and new printers with new functionality are becoming available at a rapid pace. If possible, limit your focus to printers that are in your price range. Table 1.2 lists some sources to follow that are more readable for those who do not wish to delve into highly technical details.

Make magazine (http://makezine.com) is a great source for keeping up with the latest news in the world of 3D printing, including printer recommendations. It is available digitally, in print, or as a print and digital bundle. The website also offers limited free content to nonsubscribers.

Library Service Trends

One way to filter the volume is to focus on 3D printers in libraries. Given the wide range of pricing (from three to seven figures) and functionality of 3D printers, limiting your attention to models that are in use at other libraries is a sensible way to reduce the load.

In libraries, the focus is as much on the service as well as the technology being offered through that service. To meet the need of identifying colleagues to discuss the challenges of running 3D printing services in libraries, two discussion lists have been established.

Table 1.2. General 3D Printing Websites

GENERAL 3D PRINTING WEBSITE	WEBSITE	NOTES
3ders	http://www.3ders.org	Has "3D Printer Basics," videos, forums.
3D Print Board	http://3dprintboard.com	Large forum with many threads, including hardware, tips and tricks, and "3D Printing in Education."
3D Printing for Beginners	http://3dprintingforbeginners.com	Has printer reviews, link to download 3D Systems' buyer's guide, beginners' corner, lots more.
Make magazine	http://makezine.com	In print or digital format, some content free, very readable.

The textbox includes instructions for joining the lists. Each list maintains an archive of previous posts, and newbies are encouraged to consult the archives for answers and to gain a feel for the sort of conversations held on the lists.

Key Points

Your decision to pursue a 3D service in your library will be influenced by your location, your financial and staff resources, and your vision for how best to serve your patrons.

- A 3D printing service can fit the mission of all types of libraries.
- Patrons are creating useful objects, not just trinkets, with 3D printers.
- If you do not have plans to offer a 3D printing service soon, you can refer your patrons to commercial services.

The next chapter provides some technical background about 3D printing and assessment options for selecting a 3D printer. If you are unfamiliar with 3D printing vocabulary, please skim the glossary before proceeding with subsequent chapters. Refer to the glossary early and often as you continue reading!

References

Atlantic City Free Public Library. 2015. "About @ Make@ACFPL." Atlantic City Free Public Library. Accessed August 10. http://make.acfpl.org/about/.

Beaufort [SC] County Library. 2015. "FAQ | BCL CreationStation." Beaufort County Library. Accessed August 10. https://bclcreationstation.wordpress.com/faq/.

Britton-Smedley, Laura, Meg Backus, and Thomas Gokey. 2011. "Public Libraries, 3D Printing, FabLabs and Hackerspaces." YouTube video, 10:34. Posted by Mybluheaven. April 25. https://youtu.be/HCXlJ36x-q0.

Fayetteville [NY] Free Library. 2014. "Makerspace FAQs: 21 Questions about FFL Makerspaces." Fayetteville Free Library. http://www.fflib.org/make/makerspace-faqs.

Griffey, Jason. 2012. "Absolutely Fab-Ulous." *Library Technology Reports* 48 (3): 21–24.

Harris, Jennifer, and Chris Cooper. 2015a. E-mail message to the authors. July 9.

———. 2015b. "Innovation Lab & Makerspace." *SNHU Shapiro Library [Guide]*. Southern New Hampshire University. July 13. http://libguides.snhu.edu/makerspace.

Innisfil [Ontario] Public Library. 2014. "Hacker Ethic at the Library." *The Scoop* (Aaron's blog). June 10. http://www.innisfillibrary.ca/content/hacker-ethic-library.

Jiang, Amy, and Liberty McCoy. 2015. E-mail message to the authors. July 30.

Keene [NH] Public Library. 2015. "3-D Printer." City of Keene, New Hampshire. Accessed August 10. http://www.keenepubliclibrary.org/3Dprinter.

Kent State University Libraries. 2015. "3D Printing at the SMS." Kent State University. August 7. http://libguides.library.kent.edu/3d.

MakerBot Industries LLC. 2015. "Thing Categories." Thingiverse. MakerBot Industries LLC. Accessed September 17. http://www.thingiverse.com/categories.

Makerspaces in Idaho Libraries Project. 2015. "Outcomes." Make It at the Library: Where Idaho Makers Meet. Idaho Commission for Libraries. Accessed August 10. http://libraries.idaho .gov/page/make-it-library-where-idaho-makers-meet.

Russell Gonzalez, Sara; Amy Buhler, Denise Bennett, Tara Cataldo, Michael Howell, Margeaux Johnson, Vernon Kisling, Michelle Leonard, Ann Lindell, Valrie Minson, Cliff Richmond, and Melody Royster. 2013. "3D Print Lab in the UF Libraries." University of Florida Technology Fee Proposal (unpublished). The Institutional Repository at the University of Florida. April 15. http://ufdc.ufl.edu/IR00004347/00001.

Ryan, Susan M. 2015. "Re: 3D Printing Service—General Questions." *Librarymakerspace-L Discussion List.* University of Florida. May 20. http://lists.ufl.edu/cgi-bin/wa?A0=LIBRA-RYMAKERSPACE-L.

Sacramento Public Library. 2013. "3D Printer Policy and Procedure." Sacramento Public Library. Accessed August 10. http://www.saclibrary.org/About-Us/Policies/3D-Printer-Policy-and-Procedure/.

San Diego Public Library. 2015. "Maker Lab." *Resource Guides.* San Diego Public Library. August 4. http://sandiego.communityguides.com/MakerLab.

University of Arizona Libraries. 2015. "3D Printing." Arizona Board of Regents for the University of Arizona. April 9. http://www.library.arizona.edu/services/print/3D.

Urban Libraries Council. 2015. "TechCentral MakerSpace." Urban Libraries Council. Accessed August 11. http://www.urbanlibraries.org/techcentral-makerspace-pages-330.php.

3D Printing and Printers

IN THIS CHAPTER, YOU WILL LEARN about 3D printer technology. You will find resources and factors to consider when selecting the most appropriate 3D printer for your library. Also, you will learn how to identify the best location for your printing service that maximizes safety and visibility of the printers.

What Is 3D Printing?

Three-dimensional (3D) printing is the process of building a physical object, layer by layer, from a three-dimensional digital model. This technology is considered additive manufacturing, and it has revolutionized small-scale fabrication because it allows users to create unique specialized objects that would be cost prohibitive to produce otherwise. 3D printing is not a new concept, but recent developments in technology and materials, accompanied by decreasing costs, have made printers affordable for average consumers, libraries, and educators. The accompanying printer software has matured to become more consumer friendly, opening up educational opportunities for students and users with little computer or engineering expertise. As 3D printing has exploded, its use cases have diversified, greatly moving beyond prototyping or draft printing of designs to applications including medicine, jewelry, chocolate, and pharmaceutics.

3D printing dates back to the mid-1980s with the advent of multiple types of 3D technology. Charles Hull, considered to be the inventor of 3D printing and a founder of 3D Systems, patented what is considered to be the first stereolithographic 3D printer (Hull, 1986), a technology that is described later in the chapter. Soon after, S. Scott Crump developed a new type of printer using technology that deposited layers of plastic (Crump, 1992), later trademarking the term "Fused Deposition Modeling" (FDM).

Once some of the key patents controlling 3D printing technology expired, the field opened up to engineers and tinkerers eager to explore 3D printing (Hornick and Rowland, 2013). The concept of 3D printers that can print out new 3D printers, piece by piece, was introduced by Adrian Bowyer in the mid-2000s, giving rise to the "RepRap" (replicating rapid prototyper) movement. Early versions included Bowyer's Darwin and Mendel printers and Josef Prusa's updated Mendel named the Prusa Mendel (Horvath, 2014). Not only were the printers designed with components that could be 3D printed or easily sourced, but these inventors were eager to have others improve and expand upon the designs. The plans and digital files were shared online as open source, leading to an explosion of new designs and improvements. To understand the significant impact that the RepRap project has had upon 3D printing, consult the RepRap family tree (RepRap.org, 2014). The tree summarizes six years of development from 2006 to 2012 and has over four hundred projects linked to the original RepRap 3D printers.

With the advent of crowdfunding sites such as Kickstarter and Indiegogo, inventors are able to appeal directly to potential buyers of 3D printers, allowing new technology and models to be brought to market more rapidly than ever possible in the past. Some popular 3D printers and related technology that first appeared as a crowdfunding project include the Formlabs Form1, LulzBot TAZ 3, Printrbot Simple, 3Doodler, and the Filastruder. The field of 3D printing is moving extremely quickly, and it can be tempting to purchase the newest and lowest-cost 3D printer offered in a crowdfunding campaign. However, since many funded projects never come to fruition, there is no guarantee that you will receive a product for your payment or that it will be supported and work as advertised.

For more information describing the history and basics of 3D printing, along with some deeper explanations of the technology, consult the reading list in the textbox.

3D Printing Technology

There is a wide range of types of printers, and they can cost from $350 to six or seven figures. However, the most common types for libraries and educational settings are **fused filament fabrication (FFF)** and **stereolithography (SLA)**. Other technologies include **sintering powdered material with lasers (SLS)** or even layers of paper glued together (Mcor Technologies, 2015), but these commercial-level printers and materials are cost prohibitive and their high resolution levels are unnecessary for most consumer uses. The scope of this book will focus on FFF- and SLA-type printers, and primarily FFF printers, due to their overwhelming popularity in schools and libraries.

🌀 Websites

3D Printer. "What Is 3D Printing? An Overview." http://www.3dprinter.net/reference/what-is-3d-printing.

3D Printing Industry. "3D Printing Basics: The Free Beginner's Guide." http://3dprintingindustry.com/3d-printing-basics-free-beginners-guide/.

3ders.org. "Printing Basics." http://www.3ders.org/3d-printing-basics.html.

Crawford, Stephanie. "How 3-D Printing Works." HowStuffWorks. http://computer.howstuffworks.com/3-d-printing.htm.

Ginsberg, Sharona. "Libraries & Maker Culture: A Resource Guide." http://library-maker-culture.weebly.com.

Instructables. "Introduction to 3D Printing." Autodesk Inc. http://www.instructables.com/id/3D-Printing-1/.

🌀 Videos

Harouni, Lisa. 2011. "A Printer on 3D Printing." TED Conferences LLC., video, 14:42. November. https://www.ted.com/talks/lisa_harouni_a_primer_on_3d_printing.

Public Broadcasting Service (PBS). 2013. "Will 3D Printing Change the World?" PBS Video, 7:00. March 1. http://video.pbs.org/video/2339671486/.

🌀 Books

Burke, John J. 2014. *Makerspaces: A Practical Guide for Librarians.* Lanham, MD: Rowman & Littlefield.

France, Anna K. 2014. *Make: 3D Printing.* Sebastopol, CA: Maker Media.

Hausman, Kalani K., and Richard Horne. 2014. *3D Printing for Dummies.* Hoboken, NJ: Wiley.

Horvath, Joan. 2014. *Mastering 3D Printing.* New York: Apress.

Fused Filament Fabrication Printers and Material

The most common type of consumer-level 3D printer is the fused filament fabrication (FFF) printer. FFF-type printers extrude thin layers of plastic onto a flat build plate to slowly build a 3D model from the bottom layer upward. This technology can be easily described as a robotic glue gun, building models layer by layer, and is also known as Fused Deposition Modeling (FDM), although that term is trademarked (Stratasys Ltd., 2015; USPTO, 2011). The plastic filament, 1.75 mm or 3 mm in diameter, is wound on a spool, similar to sewing thread, and is pulled into an extruder nozzle that heats the filament. The temperature required to extrude the material depends upon its composition, and temperature variations are possible within the same type of plastic sold by different vendors.

The printer follows **G-code** instructions calculated by software that specifies the extruder nozzle travel path and amount of plastic to extrude at each point. This path is described using Cartesian coordinates (**x,y,z**) with the layer deposited in the x-y plane and the distance between the plate and nozzle incrementing slightly for each subsequent layer. Figure 2.1 illustrates the FFF-style printer geometry. The printer's resolution limit is determined by how finely the plastic is extruded and the precision and calibration of the motors. The minimum layer thickness for many FFF-type printers is 100 microns; most patron models will not require that level of resolution.

One major challenge with printing models using an FFF-style printer is that each layer must be deposited onto either the build plate or a previous layer, thus requiring support material to assist building models with overhangs and protrusions. This support material is either made of the same plastic as the model and designed to snap off or, if the printer has multiple extruders, can be made using another type of filament that is dissolvable in water or a chemical solution. Depending on the model geometry, some models require a great deal of support material and thus are not feasible or too challenging to print using this technology. Figure 2.2 illustrates a 3D model of a helicopter that requires extensive support material to buttress the blades and tail. Chapter 4 provides insight on how to best design 3D models and orient them on the printer's build plate to minimize the need for support material.

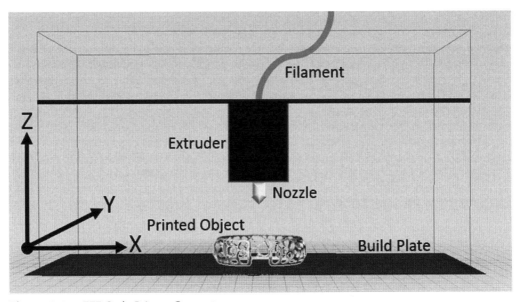

Figure 2.1. FFF-Style Printer Geometry

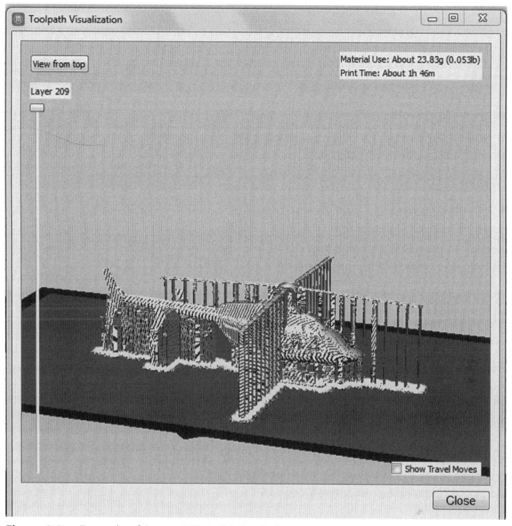

Figure 2.2. Example of Support Material Needed

FFF Filament Types

Many types of filament material can be used in an FFF-type 3D printer, but there are significant differences in their material properties, and not all 3D printers can print all types of filaments. The filament discussion is opened here because in many libraries, its environmental properties may govern which printers can be considered for purchasing.

The two most common filaments are made from polylactic acid (PLA) and acrylonitrile butadiene styrene (ABS). ABS is familiar to most users as the plastic used for LEGOs and is known for its strength and smoothness. It is petroleum based and emits a strong smell that some users find unpleasant and throat clogging, especially in closed spaces. In contrast, PLA is derived from plant starch and thus produces a fainter, sweeter smelling odor as it prints. PLA is also extruded at a lower temperature than ABS and does not require a heated build plate for the model to adhere to the plate. PLA itself is generally considered "safe" for purposes such as contact with food, but cautious users must consider any binders or dyes added during the creation of filament. PLA is more brittle than ABS and softens at 50°C/122°F, making it unsuitable for models exposed to high temperatures including hot water or even a hot car. These properties and more details are available on several websites (MakerGeeks, 2015; MatterHackers Inc., 2015; 3dprintingforbeginners, 2013).

Beyond these two major filament materials, filament makers and materials scientists are actively developing new types of material for 3D FFF printing. These specialty or exotic filaments have unique material properties that allow makers to print objects that would be impossible using standard ABS or PLA filament. Table 2.1 compares the properties of different filament types. The data and properties in table 2.1 come from MakerGeeks (2015), MatterHackers (2015), and 3dprintingforbeginners.com (2013).

Filament Extruders

As your printing service grows in popularity and you build up a pile of scraps, the idea of melting and extruding scraps to re-form filament becomes very intriguing. This concept is also appealing for libraries that are trying to be "green" with sustainability initiatives (Kreiger et al., 2014). It is possible to make your own filament, but there are caveats. "Filament that is thicker in some places than in others is difficult for an extruder to grip.

Table 2.1. Comparison of FFF Filament Material Properties

NAME	EXTRUSION TEMPERATURE	POSITIVES	NEGATIVES	NOTES
ABS (Acrylonitrile Butadiene Styrene)	210–250°C	Strong, flexible, smooth surface, withstands higher temperatures, easy to sand and paint	Strong odor, warps easily, requires heated build plate, may require fume hood, not biodegradable	Same plastic as LEGOs
PLA (Polylactic Acid)	190–240°C	Low warping, doesn't require heated build plate, recyclable, sweet odor	Less flexible, can be brittle, softens at 60°C	
HIPS (High Impact Polystyrene)	230–265°C	Dissolves only in limonene-D so good for support material		Preferred material for LulzBot 3D printers
PET(E/T/G) (Polyethylene Terephthalate)	210–250°C	Recyclable, strong, transparent surface		Same plastic as soda bottles, preferred filament for Cube's Ekocycle
PVA (Polyvinyl Alcohol)	190–210°C	Dissolvable in water	Filament susceptible to humidity	Useful as support material
Nylon	245°C	Absorbs color well (see Horne, 2013)	Filament susceptible to humidity	
NinjaFlex	210–240°C	Prints are flexible	Expensive, adhesion to build plate can be difficult	
Lay Brick and Lay Wood	Varies by intended texture effect	Achieve look of brick or wood using FFF printer	Particles can damage extruder	Mixture of PLA and brick or pine shavings

It can also result in poor print quality, with some sections too thin and some too thick" (MakerBot Industries LLC, 2015). In addition, a printer's warranty may be voided if nonapproved filament is used.

Filament is created by melting either raw virgin plastic pellets or a combination of ground recycled plastic with pellets. Pellets are poured into a hopper, melted, and extruded through a nozzle sized for the correct diameter of your 3D printer's extruder head. As the filament is extruded, it is wound with adequate tension onto a spool holder, taking care not to tangle or bend the newly formed filament.

Advantages of extruding your own filament are that you can control the color of filament by varying the shade and amount of colorant added to the mixture. You will also save money since the price of plastic pellets is significantly cheaper than the equivalent mass of filament. However, extruding filament requires a commitment of labor and attention to detail in order to produce enough high-quality filament for your printing needs. Several filament makers are available for purchase, and the two currently best known are the Filastruder (http://www.filastruder.com) and the Filabot (http://www.filabot.com) machines. The Filastruder extrudes at 9–18 inches/minute and the Filabot Original extrudes at 6–20 inches/minute. A one-kilogram spool of PLA contains approximately 330 meters of filament so, at these rates, it will take eleven to thirty-six continuous hours to extrude a similar amount of recycled filament (ToyBuilder Labs, 2013).

3dprintingforbeginners.com (2014) provides a conversation on making your own filament. At the present time, given the low cost of filament and the time and labor it takes to extrude a one-kilogram spool, making your own filament is likely not a desirable option for most libraries. The dream of accepting recyclables to convert to usable filament isn't a reality yet.

You or your patrons may find other uses for your filament scraps. They can be melted down and poured into solid molds, or used to "glue" or repair other 3D printed parts that didn't turn out as expected.

Stereolithography (SLA) 3D Printers

As noted in the history of 3D printing, one of the earliest types of 3D printing used stereolithography (SLA) to print models. In contrast to FFF printers that deposit a layer of material, SLA printers build models by hardening liquid photopolymer (resin) using an ultraviolet wavelength laser. The laser is focused using a mirror, tracing a path that describes a layer. Once the resin is hardened, the printer advances to the next layer. Figure 2.3 illustrates this process. SLA printing still requires support material for overhanging components that must be removed manually.

SLA-printed models differ from FFF-type models in resolution potential and ability to print at small scale. SLA printers are capable of much higher resolution than FFF printers, with the Formlabs Form1's resolution limit at 25 microns compared to 100 microns for most FFF printers (http://formlabs.com/products/form-1-plus/). Depending on the type of SLA printer, there are many formulations of the resin available in various shades and material properties. Most resin models are extremely strong and possess highly smooth surfaces.

These properties make SLA printers very desirable for makers who need high resolution or the ability to print very small objects, such as for jewelry or small game pieces. However, uncured resin must be stored out of sunlight, with a shelf life of about one year, and disposed of properly. Users are also recommended to wear gloves and protective

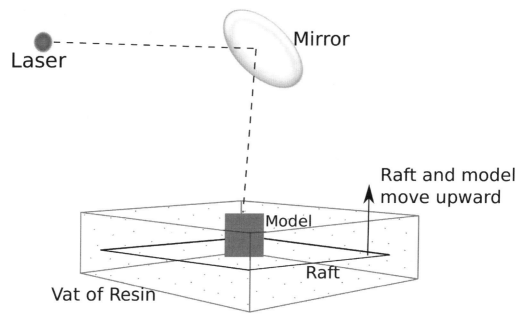

Figure 2.3. Stereolithography (SLA) 3D Printing Process

eyewear when handling the resin. To find material properties and recommended handling practices, consult material safety data sheets such as the example at http://www.maker-juice.com/docs/SubSF-MSDS.pdf.

Location

An early issue to resolve is where the 3D printing service should be located within the library. Location parameters may dictate the selection of printer features and therefore of specific models. If you have already selected a printer, your printer's features and size may constrain the location.

Size and Space

The most obvious consideration is the size of the 3D printer and how much surface space it needs to print successfully. 3D printers should be placed on a surface that is relatively free from vibrations and offers working space for changing filament and removing models. Ideally, you should have enough room to access all sides of the printer. Some 3D printers must remain tethered to a computer to both calibrate the printer and transfer models to be printed. If you choose a tethered model, you will have to accommodate the computer's footprint as well as that of the printer. Other printers can print remotely using SD cards or even over a wireless network. Be sure to consider how users and staff will transfer files when considering the size of and access to the 3D printer location.

Most printers are accompanied by accessories, such as printer tools and supplies and support removal tools, and these all require storage near the printer. Additionally, filament spools should be kept in a location away from moisture and dust, ideally in a cabinet or spool rack to facilitate easy filament changes (3dprintingforbeginners.com, 2015). Resin for an SLA printer must be kept at or below room temperature, optimally away from strong light.

If an FFF-type 3D printer is used, be prepared to manage the filament debris that is inevitable after each model print. Most printers lay down a test line of plastic to clear the model and this, along with model supports and rafting, will accumulate near the printers. If the filament is PLA, these scraps may be recycled, but other materials will need to be placed in the trash. Filament bits will adhere to carpet, so smooth flooring or a carpet covering may be preferable. One 3D printer's mess is fairly easy to contain, similar to an area where you may supply a pencil sharpener, hole punch, and stapler. If you expand to a fleet of 3D printers or a fuller makerspace, you will need to devise a method for controlling and cleaning the noise and mess.

If an SLA-type printer is selected, allow counter space for postprocessing of the models and facilities to dispose of waste resin. If possible, locating the printer and processing area near a sink or drain will assist ease of cleanup. Make sure this processing function will be permitted by your institution.

Furniture

You may discover a desire for appropriate furniture to support the service. Each library will have different needs, depending on the location and the type of service (staff mediated or patron initiated; see chapter 6) chosen. Work tables must be sturdy enough to support the printer without it wobbling while it is printing. Additional tables and seating may be desired to accommodate personal space, collaborative space, and computer space. Storage for tools and accessories is critical, and some may need to be lockable. Ventilation equipment such as a laboratory extractor or a modular filter system may be recommended or deemed necessary (see "Environmental Health and Safety," later in this chapter) by your institution. Signage to indicate where supplies are located will be necessary.

Staff and Patron Access

3D printers should be placed within easy access of staff to monitor print quality and facilitate swapping of models. If patrons are responsible for operating the printers, staff should check periodically on the printer status to ensure that the printing hasn't gone awry. Common printer errors occur frequently with 3D printing (see chapter 10), and to avoid wasting time or money, or damaging the printer, staff ideally should be in view of the printers to cancel a print job early in a failure.

If staff are not in nearby or visible range of the 3D printers, consider setting up remote monitoring using streaming webcams. Some 3D printing manufacturers are including cameras built into the printer housing, but webcams can be mounted to allow staff and patrons to monitor jobs in real time. Remote viewing may be especially important if the 3D printers operate during a time when the library is closed, an issue that must be addressed to accommodate jobs that run longer than originally estimated.

You will find marketing value in placing the 3D printers in a highly visible location since patrons will then be able to watch models print. Many patrons may have never seen a 3D printer in person, and an opportunity to view can be the first step to the road to printing and even 3D modeling. Place informational literature near the printers along with workshop schedules or related services to reach interested patrons. Another promotional opportunity is to create a gallery of objects, including failures, that can illustrate to patrons the potential for 3D printing and what modeling pitfalls to avoid. The gallery may be located near the printer or the service point or both.

Environmental Health and Safety

One of the first questions often asked about 3D printing is the health impact of fumes emanating from the 3D printers. Very few studies have been published that examine the toxicity of particulate emissions from 3D printers, but more are forthcoming, especially as consumer 3D printers become ubiquitous. The most well known is a study by Stephens et al. (2013) who analyzed air quality surrounding printers printing PLA and ABS and discovered that toxicity depends on the type of material, printer enclosure, and print volume. They found that printers using ABS emitted ultrafine particles (UFP) a magnitude higher than when printing with PLA; however, both were deemed as high-emitters. The significance of these results is unclear since other particle sources, such as laser printers, candles, and cooking over stoves, have similar levels of UFP emission.

Depending on where the 3D printers will be located, printer noise should be considered in order to minimize impact upon staff and patrons. Most printers individually do not produce a high level of noise, but if several printers are operating jointly, the hums and beeps may disturb nearby individuals. The sound generated by one printer may or may not reach an annoying level, depending on the printer, its location, and the times when it is in use. But if you add a second printer (or more), location placement and noise dampening solutions will become critically important for your staff and for your patrons. Makes and models vary in their auditory output, so pay attention to their volume if the noise level is an important element on your checklist.

Parts of the 3D printers do become extremely hot so, if the printer is in an area frequented by patrons (especially children), take precautions to keep the printer out of reach. The extruder is the most dangerous part of the printer, reaching temperatures of 260°C/500°F. Even a heated build plate can cause injury since the standard plate temperatures can range from 45°C/113°F (PLA) to 120°C/248°F (ABS).

If environmental safety is a concern, you can take several steps to mitigate risk. One of the simplest is to surround the 3D printer in a transparent enclosure or "cake topper" made of Plexiglas, glass, or acrylic that will trap the fumes and minimize the printer noise. Fumes can be removed more thoroughly with a professional ventilation system. Most libraries will likely not have such a system in place, and thus this highlights the importance of clearing specific 3D printers and materials through each library's appropriate channels before purchase. If chemical solutions are used postprinting, appropriate waste removal procedures must be followed and a staff member may need to be trained to correctly handle these materials.

3D printers work best in climates that are controlled for temperature, humidity, and drafts. The print jobs would rather be too warm than too cool, and you should avoid temperature fluctuations to prevent rapid or irregular cooling, which leads to warping and problematic removal. "Cake toppers" may help with climate control if you don't have a cozy location for the printer.

Each manufacturer should have a material safety data sheet available, so consult these if there are questions about toxicity and safety procedures. Because so few studies concerning the toxicity of 3D printer emissions have been published, 3D printers and setups allowed in one library may be banned in another. If you prefer a printer or filament that is blacklisted by your institution, consult other libraries for solutions and examples on how to safely print.

⑥ Selection of Printer and Accessories

Hundreds of 3D printers are currently on the market, and selecting the best one for your library can initially appear to be a daunting process. However, by identifying the key parameters that are important for your library and your patrons and by consulting reputable sources for recommendations, you will be able to narrow down this list to printers that meet your requirements.

Factors to consider in 3D printer selection include:

- Printer dimensions
- Assembled or DIY kit
- Printer filament material type
- Printer features
- Warranties
- Availability of parts
- Nearby 3D printer labs
- Software requirements

Printer Dimensions

When discussing size, you need to consider both the physical size of the printer and the build size, the dimensions of the largest object a printer can build. 3D printers vary widely in physical size and weight, from floor-size models to portable printers that are similar in dimension to a toaster oven. For example, the Afinia H480 3D printer has a footprint of 9.64 × 10.23 × 13.78" and weighs less than 11 lbs. This is in contrast to Stratasys' UPrint SE+ with two material bays that has a much larger footprint of 25 × 26 × 37" and weighs 206 lbs. You will need to identify a flat surface capable of holding the printer's weight along with space to hold tools and other equipment.

The physical size and weight are a factor when determining where in the library to locate your printer and when assessing portability if the printer will travel to classes or workshops. If you plan to transport the printer, select one that can be easily lifted by a single person and ideally can be stored without disassembly and within a sturdy case. For example, the Printrbot Simple Metal fits snugly within a Pelican 1620 case, with room for cords and tools. Another option is the Ultimaker 2 Go, designed to be portable with its own custom case.

The maximum build size of a printed model correlates roughly with the printer footprint, and is determined by the size of the build plate and the distance from the plate to the upper extreme of the extruder nozzle. For the printers used as examples above, the build sizes aren't quite as dissimilar, with the H480 capable of printing models approximately 5 inches cubed and the UPrint SE+ printing up to 8 × 8 × 6 inches. If a model is too large for the printer's build size, patrons will either have to rescale it or slice it into parts to glue together after printing. Most printers cannot be modified later to increase the build volume, so query your patrons to discover what models they plan to print to ensure that your printer is large enough for patron needs.

The dimensions and weight of 3D printers are generally listed on the manufacturers' websites. You may need to hunt to find the technical specifications.

Assembled or DIY Kit

Many 3D printers can be purchased either in a do-it-yourself (DIY) kit form or, for a higher cost, fully assembled. It can be tempting to save money by purchasing a kit, but assembling the printer is not trivial and will require time, tools, and a basic comfort level with electronics. Before purchasing a kit, read through the assembly guide and watch any tutorial videos available to ensure that this is a project worth undertaking. Your library may have patrons with experience building 3D printers or similar hardware, so be sure to tap into your local community for assistance if you decide to purchase a kit.

Filament Material Type

As described earlier, consumers have a wide variety of materials from which to select for 3D printing. Some printers are limited to only one type of material because they are only capable of heating to a narrow range of temperatures or lack a heated bed, which is necessary for some plastics such as ABS. Other printers can heat to a wider range of temperatures and thus can accommodate a variety of materials. Note that some 3D printers are restricted to proprietary filament and cannot accept another brand of material. If you choose such a printer, your purchase and color options will be limited. SLA-type printers can only use liquid resin, although it does come in several different shades.

In addition to different types of materials, most manufacturers offer a wide palette of colors. Decide if you will only stock a single shade or expand and offer a rainbow of colors. White is a good basic color to provide, since it is widely available from manufacturers and patrons will find it easy to paint. Know that extrusion temperature may vary based upon color and manufacturer of the filament. Vendors will supply the recommended temperature to use, but it may also take experimentation to find the exact temperature value to use with your printer in the environment where it is housed. If you have filaments that require different temperatures to extrude, you will need to make sure that all staff and patrons know those values when preparing models to print using the printer software.

Two different filament diameters are available, 1.75 mm and 3.0 mm, but any given extruder can only accommodate one size. Typically there is no difference in price/kg, and color choices are similar. If you decide to purchase different models of 3D printers, strongly consider selecting printers that use the same diameter of filament. It will be one less variable to track when managing your supplies and your jobs.

If you are willing to manage multiple types of material, survey your patrons to identify what properties or colors they would prefer for their prints. You may have additional material restrictions based upon health concerns as detailed in the "Environmental Health and Safety" section above.

Printer Features

The majority of FFF 3D printers have a single extruder nozzle, which limits the printer to using a single spool of filament at a time. If you wish to print in multiple colors or print support material in a different type of filament than the main model, you will need to purchase a printer with more than one extruder. Examples of dual extruder 3D printers are the MakerBot Replicator 2X, the Leapfrog Creatr, and the Flashforge Dreamer. Cubify's CubePro line has a printer that can print three materials at a time. Stratasys's Mojo and UPrint printers allow users to print models with one filament while the support structure is printed using a material that dissolves in a separate WaveWash system. While multiextruder printers offer the potential to produce impressive models, these types of printers can prove tricky to operate and may not be the best choice for a library that does not have the staff or time to focus on solving technical issues repeatedly.

A computer with sufficient graphics capabilities (depending on printer and model file size) is required to process the STL files according to the printer's specifications. In addition, some 3D printers must be tethered to a computer to enable the file transfer of the sliced job to the printer. Other printers do not require tethering, and files can be transferred via USB, SD cards, wireless, or other options. Printers that require tethering may also require a dedicated computer or laptop. Tethering might be a deal breaker if you don't have an extra computer to dedicate, if you prefer not to keep a computer attached to your printer, or if you envision moving the printer between its working spot and training venues. Devices to eliminate tethering, such as the MatterControl Touch stand-alone controller, are being developed for several printers. A network port for the computer is preferable to permit downloading of STL files from model websites such as Thingiverse .com, shared sites such as Dropbox, or from patrons' e-mail accounts. Electrical power is required for both the printer and computer.

Warranties

Warranties vary by manufacturer, and you should avoid any printers that do not come with a warranty unless you have the time, budget, and expertise to repair the printer yourself. Purchase the longest warranty that your budget allows to ensure that your printer operates smoothly for as long as possible. Read the warranty policy closely to learn if any printer modifications or use of nonmanufacturer print material will void the contract. See chapter 10 for a discourse on the challenges of maintaining these machines.

Availability of Parts

Look for printers that have easily sourced replacement parts that are available from a well-stocked manufacturer website or other storefront that carries the necessary pieces. Take note of the prices of parts that are likely to wear out (such as belts, extruder nozzles, motors) and factor those into the overall cost of the printer. Some printers, such as LulzBot and RepRap, are open source and a subset of replacement components can even be 3D printed. Consider whether printer components are unique to that particular printer or are generic ones that can be sourced through multiple vendors.

Nearby 3D Printer Labs

If other 3D printing services are located near your area, learn what printers they have, what they like or dislike about those printers, and what functionality their users are requesting that cannot be delivered currently. Answers to these questions may inform your decision making about printer features. Choosing a printer that is identical or similar to a neighbor's can result in instant support and sharing of accessories. Selecting a printer with different functionality may help ensure that you are not directly competing with a neighbor's service and that you can offer more options to your patrons.

Communicating with other labs nearby may lead to opportunities for collaboration. At the University of Florida, there are several 3D printing labs on campus with 3D printers that are similar to those in the libraries. Librarians network with these other labs to seek advice on printer repair, to share supplies, and even to assist with printing when a printer is broken or overloaded with jobs.

Software Requirements

Some 3D printers come with proprietary software to slice the patron's model into a set of instructions that the printer will use to construct each layer of the item. For example, MakerBot printers use MakerWare, custom software that is designed to work exclusively with MakerBot printers. Other printers can use an open source software program to perform the model preparation. Popular options include Cura (maintained by Ultimaker), ReplicatorG, and Repetier-Host.

Each make (or in some cases, each model) of printer may use different software or printer settings, creating challenges for staff and users if you have more than one printer. A solution to learning and using several versions of printer software is to license an integrated software program such as Simplify 3D (http://www.simplify3d.com/) that can import, adjust, and slice jobs for a variety of printers. For additional details about supporting software, see chapter 8.

Summary of Printer Features

Choosing a 3D printer is similar to choosing a car: most models function decently; they vary greatly in size, comfort, and cost; their features are similar but are placed differently; new versions are available at regular intervals; and each purchaser has a unique set of priorities among the features available. The checklist in table 2.2 will help guide your selection process. Establish your priority for each feature, and then it will be easier to weigh the specifications and prices you find for each printer on your list.

Table 2.3 provides data about 3D printers that have been most commonly purchased by libraries as of this writing. Follow the bullets across the rows for features unique to each model. Many more comparative details are available on the manufacturers' websites. Use these specifications to start your search, and then use the worksheet in table 2.2 to further narrow your options. Keep in mind that even the most basic specifications are subject to change. The manufacturers tweak the hardware details as frequently as typical software updates. For additional printers, consult the Reddit wiki's printer chart (Reddit, 2015).

As though you were selecting a vehicle, spend ample time researching options that fit within your budget and that possess the features you prioritize as described above. Consult other libraries about their experience with the types of printers you are considering. In addition to features and functionality, critical questions include the printer's reliability and the quality of the vendor's customer support. Check the dates of any reviews carefully and, as always when evaluating opinions, consider the influence of a vendor on the review source. Printer models are upgraded continuously, and the pros and cons for an earlier version may no longer be valid or a printer may be too new to have valid critiques. Suggested resources to consult are listed in the textbox.

Table 2.2. Pick-a-Printer Checklist

FEATURE	YOUR PRIORITY: HIGH, MEDIUM, OR LOW	SPECS AND PRICING FOR PRINTER A	SPECS AND PRICING FOR PRINTER B	SPECS AND PRICING FOR PRINTER C
Total price: • printer • computer • tools and accessories				
Location and space preparation needs				
Vendor purchase limitations, if any, at your institution				
Printer footprint—space needed for printer				
Build size: dimensions of largest pieces printable				
File transfer method: SD card, USB, tethered to computer, Wi-Fi. Check for multiple options.				
Supporting software: • Costs • Ease of use				
Filament options: • Materials • Sizes of containers • Prices • Proprietary required or recommended				
Ease of use by staff or patrons: • Printer • Processing software				
Warranty options and reliability				
Service options and availability of parts				
Safety concerns: • Heated bed plates • Other hot or moving parts				
Health concerns: • Particle emissions • Noise level				
Portability, if printer will travel				

Table 2.3. Specifications for 3D Printers Common in Libraries

MODELS, MANUFACTURER, WEBSITE	FILAMENT OR RESIN	FOOTPRINT (INCHES) W X D X H	BUILD AREA (INCHES) L X W X H	SOFTWARE	ONLINE MANUALS AND GUIDES	FILE TRANSFER METHOD	WARRANTY	PRICE
Afinia • H480 • H800 http://www.afinia.com	1.75 mm • ABS • ABS, PLA	• 9.64 x 10.23 x 13.78 • 19.1 x 20.5 x 19.5	• 5.5 x 5.5 x 5.3 • 10 x 8 x 8	Afinia, download online	Guides, manuals, videos	Tethered or USB, can disconnect from computer when job starts	Yes, one year	$1299 $1899
Cubify • Cube • CubePro http://www.cubify.com	Cartridges • ABS, PLA • ABS, PLA, nylon	• 13.2 x 9.5 x 13.5 • 22.75 x 22.75 x 23.25	• 6 x 6 x 6 • 11.2 x 10.6 x 9.06	Cube, download online	Videos	Wi-Fi or USB	Yes, 90 days	$999 $2799
Flashforge • Creator Pro http://www.flashforge-usa.com	1.75 mm ABS, PLA	12.6 x 18.4 x 15	8.5 x 5.7 x 5.9	ReplicatorG download online, can bundle with Simplify3D	User manual	SD card or USB	Yes, 6 or 12 months	$1349
Formlabs • Form 1+ http://formlabs.com	Resin	12 x 11 x 18	4.9 x 4.9 x 6.5	PreForm, download online	Guides, manual, videos	Tethered, can disconnect when file uploaded	Yes, one year	$3299
LulzBot • TAZ Mini • TAZ 5 http://www.lulzbot.com	3mm HIPS (preferred), ABS, PLA, lots more	• 17.1 x 13.4 x 15.2 • 26.8 x 20.5 x 20.3	• 6 x 6 x 6.2 • 11.7 x 10.8 x 9.8	Comes with Cura LulzBot, download online. Compatible with OctoPrint, BotQueue, Slic3r, Printrun, MatterControl, more	Quickstart guides, manuals, support	• USB connection • SD card	One year, extended option	$1350 $2200

Printer / URL	Filament	Dimensions	Software	Build volume	Support	Connectivity	Warranty	Price
MakerBot • Mini • Replicator • Z18 http://www.makerbot.com	1.75 mm PLA		Comes with MakerBot, download online	• 3.9 x 3.9 x 4.9 • 9.9 x 7.8 x 5.9 • 11.8 x 12 x 18	Guides, manuals, videos, more	• Tethered USB, Wi-Fi • USB, Wi-Fi, Ethernet • USB, Wi-Fi, Ethernet	One year	$1375 $2899 $6499
MakerGear • MakerGear M2 http://www.makergear.com	1.75 mm PLA		Can use open source (such as Slic3r),recommends Simplify3D	8 x 10 x 8	Videos, forum	Tethered or buy LCD controller	6 months	$1775
PolyPrinter • PolyPrinter 229 • PolyPrinter 508 http://www.polyprinter.com	1.75 mm ABS	• 25 x 16 x 8 • 27.5 x 25 x 18	Download online	• 9 x 9 x 9 • 20 x 9 x 9	Guides, manual, FAQs	Tethered	6 or 12 months	$2395 $4495
Printrbot • Play • Simple • Plus http://printrbot.com	1.75 mm • PLA • PLA • ABS, PLA, flexible	• 15 x 10 x 17 • 18 x 17 x 13 • 21 x 20 x 22	Recommends open source Cura or Repetier	• 4 x 4 x 5 • 6 x 6 x 6 • 10 x 10 x 10	Guides, FAQ, knowledge base	Tethered	90 days	$399 $599 $1199
Stratasys • Mojo • uPrint SE http://www.stratasys.com Needs added WaveWash to clean supports	ABSplus	• 25 x 18 x 21 • 25 x 26 x 31	Print Wizard CatalystEX	• 5 x 5 x 5 • 8 x 6 x 6	User guides	Tethered to Windows-compatible		Contact for quotes
Ultimaker • Ultimaker 2 Go • Ultimaker 2 • Ultimaker 2 Extended http://ultimaker.com	3 mm ABS, PLA	• 10.2 x 9.8 x 11.3 • 14 x 13.5 x 15.3 • 14 x 13.7 x 19.2	Cura, free download	• 4.7 x 4.7 x 4.5 • 8.8 x 8.8 x 8 • 8.8 x 8.8 x 12	Knowledge base, troubleshooting guides	SD card	One year	€1195 €1895 €2495

> ### 3D PRINTER REVIEW RESOURCES
>
> New printer companies and models come and go at an alarming rate. For reviews and announcements of reputable options, consult these sources:
>
> - *Make* magazine's annual review issue: http://makezine.com
> - Company support forums, once you've identified brands of interest
> - Reddit's 3D Printing community: http://www.reddit.com/r/3dprinting
> - Amazon.com—see the most popular and currently available models, and read the reviews
> - 3ders: http://www.3ders.org
> - 3D Print Board: http://3dprintboard.com
>
> Join a librarians' discussion list:
>
> - LIBRARYMAKERSPACE-L@lists.ufl.edu. Send a message to listserv@lists.ufl.edu. In body of message: subscribe librarymakerspace-l <your name>
> - LITA's (American Library Association / Library and Information Technology Association) 3D Printing Interest Group mailing list. To join, go to http://lists.ala.org/sympa/info/lita-3d and follow the instructions.

Key Points

Choosing a printer seems like a daunting task, but your budget, location, and environmental factors may limit your options quickly.

- Among a wide variety of 3D printers available, libraries tend to choose those that are relatively inexpensive to purchase and operate.
- The type of filament that you are permitted to use for environmental reasons may limit the models of printers that you can consider purchasing.
- Keep in mind that choosing a 3D printer is like choosing a car. Many options will work well, but your space and functional requirements are unique. Prioritize your needed and desired features, then read the reviews and consult current owners of your short-list choices. Don't rely solely on the reviews of older models in the line.
- Choose a location for your printer that ensures safety for patrons, security and visibility for the printer, and easy access for staff.
- Mitigate environmental risks to ensure a safe printing service for both your printer and patrons.

Before you print anything, you need to obtain a model. In the next chapter, you will discover how to locate and how to create a 3D digital model suitable for 3D printing.

References

3dprintingforbeginners.com. 2013. "What Material Should I Use for 3D Printing?" *Materials* (blog). February 10. http://3dprintingforbeginners.com/filamentprimer/.

———. 2014. "How to Make DIY Filament for Your 3D Printer." *Beginners Corner* (blog). May 22. http://3dprintingforbeginners.com/how-to-make-diy-filament-for-your-3d-printer/.

———. 2015. "How to Store 3D Printing Filament—A Simple, Quick and Cheap Way to Keep Your Filament Dry." *Materials* (blog). January 14. http://3dprintingforbeginners.com/how-to-store-3d-printing-filament/.

Crump, S. Scott. 1992. "Apparatus and Method for Creating Three-Dimensional Objects." US Patent 5,121,329. Filed October 30, 1989, and issued June 9, 1992. http://www.google.com/patents/US5121329.

Horne, Richard. 2013. "3D Printing with Nylon 618 Filament in Tie-Dye Colours." *Reprap Development and Further Adventures in DIY 3D Printing* (blog). April 11. http://richrap.blogspot.co.uk/2013/04/3d-printing-with-nylon-618-filament-in.html.

Hornick, John, and Dan Rowland. 2013. "Many 3D Printing Patents Are Expiring Soon: Here's a Round Up & Overview of Them." *3D Printing Industry* (blog). December 29. http://3dprintingindustry.com/2013/12/29/many-3d-printing-patents-expiring-soon-heres-round-overview/.

Horvath, Joan. 2014. *Mastering 3D Printing.* New York: Apress.

Hull, Charles. 1986. "Apparatus for Production of Three-Dimensional Objects by Stereolithography." US Patent 4,575,330. Filed August 8, 1984 and issued March 11, 1986. http://www.google.com/patents/US4575330.

Kreiger, M. A., M. L. Mulder, A. G. Glover, and J. M. Pearce. 2014. "Life Cycle Analysis of Distributed Recycling of Post-Consumer High Density Polyethylene for 3-D Printing Filament." *Journal of Cleaner Production* 70: 90–96.

MakerBot Industries LLC. 2015. "Filament." Support: 3D Printing Basics. MakerBot Industries LLC. https://www.makerbot.com/support/new/3D_Printing/Knowledge_Base/Filament.

MakerGeeks. 2015. "PLA Filament, ABS Filament & 3D Printing Supplies at MakerGeeks.com." MakerGeeks.com. Accessed August 10. http://www.makergeeks.com/.

MatterHackers Inc. 2015. "Home." Accessed August 10. MatterHackers Inc. http://www.matterhackers.com.

Mcor Technologies. 2015. "3D Printing and Rapid Prototyping." Mcor Technologies Ltd. Accessed August 10. http://mcortechnologies.com.

Reddit. 2015. "Printerchart." *3D Printing* (wiki). August 14. http://www.reddit.com/r/3Dprinting/wiki/printerchart.

RepRap.org. 2014. "RepRap Family Tree." *RepRap* (wiki). January 12. http://reprap.org/wiki/RepRap_Family_Tree.

Stephens, Brent, Parham Azimi, Zeineb El Orch, and Tiffanie Ramos. 2013. "Ultrafine Particle Emissions from Desktop 3D Printers." *Atmospheric Environment* 79: 334–39.

Stratasys Ltd. 2015. "FDM Technology." Stratasys Ltd. Accessed August 10. http://www.stratasys.com/3d-printers/technologies/fdm-technology.

ToyBuilder Labs. 2013. "Filament Volume and Length." *ToyBuilder Labs Blog.* ToyBuilder Labs. July 13. http://www.toybuilderlabs.com/blogs/news/13053117-filament-volume-and-length.

USPTO (United States Patent and Trademark Office). 2011. "Mark: FDM." December 7. http://tsdr.uspto.gov/#caseNumber=74133656&caseType=SERIAL_NO&searchType=status Search.

Before You Can Print: Getting a Model

WHEN PEOPLE THINK OF 3D PRINTING, many focus solely on using a 3D printer, but the printing is really the last stage in a potentially lengthy process. To use a cake baking analogy, the printing part is similar to putting the pan in the oven. It's fun to watch (if you have time and patience), you must adjust for the peculiarities of your machine, and you must maintain your equipment. However, the creative functions have already occurred before that stage. In baking a cake, you have already previously decided on the size, flavor, and mix-ins, and you will get the best results if you use the proper techniques for creating the batter and preparing the pan before you transfer the batter. Likewise in 3D printing, you must first select or create your model, choose the appropriate settings, and process the file before you transfer it to the prepared printer. This chapter discusses the creative steps that precede the function of loading a file into the 3D printer.

To print a physical object, you first need a digital file that describes the object's surface in three dimensions. For 3D printers with a single extruder, color and type of material does not need to be specified. However, for multiple material printers (with two or more extruders), an additional component describing the type or color of material is necessary.

There are three ways to obtain a 3D model: download a model from an online repository, scan an object to create a 3D model, or design a model using modeling software. Staff and patrons new to 3D printing are encouraged to first explore the wide variety of free 3D models available online. Not all 3D models print or are suitable for your particular type of printer (see chapter 4 for guidance on assessing 3D models) and you will gain a great deal of knowledge by trial and error with printing. Once a user gains experience with understanding how the user's 3D printer works, then it is time to begin modifying existing models, creating new ones, and scanning physical objects.

The most common file formats for 3D printer slicing software are the STL and OBJ formats. The acronym STL is defined in different places as "STereoLithographic," "Standard Tessellation Language," and "Standard Triangle Language" (Grimm, 2004). The STL file format models a 3D object by drawing a surface mesh layer using connected triangles (see figure 3.1). The object (OBJ) file format is also considered a universal 3D file format and describes 3D models using polygons (Singh, 2015).

Many other possible formats for 3D models exist, and table 3.1 lists some of the most common file extensions along with their associated software. Knowing what software a patron used to create the model can help library staff debug errors in the model or suggest alternate software to convert the file to a more standard type.

Finding Existing Models

The easiest way to print a 3D object is to find a model that has already been designed and made available for download. Some printer manufacturers have created or sponsored online repositories, such as Thingiverse.com by MakerBot and Youmagine.com by Ulti-maker. Museums, notably the Smithsonian, are scanning some objects in their collections and making these models available both for fun and for educational purposes. Teachers will find these especially useful when combined with curricular aids for using the result-ing 3D printed objects in the classroom as described in chapter 1. U.S. federal agencies have been tasked to make their scientific collections more available to the public (Hol-dren, 2014), and "digital 3D models" are encouraged as a viable option, appropriately in Section 3(d) of the memorandum. Relevant agencies are beginning to develop websites that house 3D models, and perhaps an overarching index will soon be provided as well.

While the large and growing number of 3D models available through these various repositories enable users to quickly print a vast array of objects, it is important to keep in mind that the quality and printability of models found in repositories is not guaranteed. Museums and other single-source repositories in which the models have been created solely by the site owner tend to offer consistent and error-free quality 3D models; how-ever, in some cases these models are designed for visualization and are not suitable for 3D

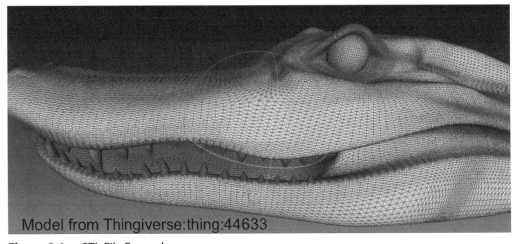
Model from Thingiverse:thing:44633

Figure 3.1. STL File Example

Table 3.1. Common 3D File Formats, Abbreviations, and Software

FILE EXTENSION		ASSOCIATED SOFTWARE & NOTES
.3ds		Autodesk software format (www.autodesk.com)
.blend		Blender (www.blender.org)
.dae	Digital Asset Exchange	3D image files based on the COLLADA format (www.khronos.org/collada/)
.dwf	Design Web Format	Autodesk software format (www.autodesk.com)
.fbx	Filmbox	Autodesk software format (www.autodesk.com)
.lwo	Lightwave Object	Lightwave (www.lightwave3d.com)
.obj	Object	Universal file type (Bourke, 2015a)
.ply	Polygon File Format	Common to many software (Bourke, 2015b)
.skp	Sketchup	Sketchup (www.sketchup.com)
.shp	Shape	Developed by ESRI for storing spatial data, can be converted to .stl file to 3D print (McCune, 2014)
.stl	Standard Tessellation Language	Universal file type, common in most software, also known as stereolithographic (Bourke, 3 1999)
.thing		MakerBot (www.makerbot.com)
.vrml	Virtual Reality Modeling Language	Created as web-based 3D format in 1995 (Web3D Consortium, 2015)
.x3d	X3D	XML encoding of VRML format (Web3D Consortium, 2015)

printing (see chapter 4 for details about why some models are not printable). In contrast, models found in repositories where users upload their own files will vary greatly in quality and printability. In addition to some files not being designed for 3D printing, files created by inexperienced modelers may be flawed, and models may be optimized for specific printers. Most online repositories allow all users to upload models without restrictions or verification, and thus models should be examined carefully and critically for quality before printing. Most librarians are very familiar with these challenges since this is analogous to using information from the crowdsourced *Wikipedia*.

Some uploaded files have been created in modeling software but have never been printed, so their creators may be unaware that their models cannot be printed successfully. Caution your staff and patrons to look for photos of completed objects that have been 3D printed as a first measure of printability. In addition to printability, staff and patrons must observe whether there is a charge to download a model from a repository, or whether the design is licensed under a Creative Commons (http://creativecommons.org) or other license. While noting whether a designer is protecting a model, library staff may need to observe whether a model is a facsimile of a design, such as a cartoon character, that is likely to be protected from reproduction under a license or trademark. Your library's policy may restrict the printing, especially in quantity of protected models. See chapter 7 for more details on establishing a library policy that balances freedom of expression and intellectual property rights.

Some repositories give the appearance that their models are downloadable, but instead they are primarily a storefront for customers to request that an item be 3D printed and shipped. Other repositories contain 3D models that are meant for other 3D applications, such as computer animation, video games, and architectural renderings. Library staff should guide patrons in possession of unsuitable models to search for similar objects in one of the key repositories listed below.

Key Repositories

Selected repositories to search for 3D models are described below. They are summarized for easy skimming along with links to their websites in table 3.2.

Starter Sites

Thingiverse.com, affiliated with MakerBot, is perhaps the best-known online repository. Users can either search using keywords or browse general categories. Registered users can upload and save models, and make them available for anyone to download. Many files can also be modified using embedded customizer features. Be aware that the database also contains files for laser cutting, plans for building 3D tools, and software. Thingiverse frequently hosts contests for users to submit models that fit a theme or set of parameters, and features seasonal or holiday models in its banner.

YouMagine, affiliated with Ultimaker and a competitor to Thingiverse, supports open source creation and use of open source software tools. Features include the ability to upload and comment upon models and an idea section for users to collaborate with designers. You-Magine spearheaded a new open source license for 3D models, called 3DPL (Peels, 2015).

Smithsonian X3D makes available for 3D printing the STL files of many objects and specimens digitized from its collection. The website includes a link for educators that describes pilot projects and case studies for using 3D models in the classroom. Teachers are invited to submit their success stories.

Table 3.2. Online Repositories of 3D Models for Printing

REPOSITORY	AFFILIATION (IF ANY)	DESCRIPTION & NOTES
3Dagogo http://www.3dagogo.com/	AstroPrint https://astroprint.com	Models verified to be printable. Has a communal storefront. Personal, business, and commercial level licenses may available for each model. Some free, some cost to download.
3D Warehouse http://3dwarehouse.sketchup.com	Trimble & SketchUp	Large repository of 3D objects, not all designed for printing.
Bld3r http://www.bld3r.com/		Open source, can host models or link to models hosted elsewhere, share tutorials.
MakerHome http://makerhome.blogspot.com and http://mathgrrl.com/hacktastic/	Dr. Laura Taalman	365 (more or less) models on MakerHome. Most designs are math oriented. Completed project of one model per day for a year. Blog continued on Hacktastic.
NIH 3D Print Exchange http://3dprint.nih.gov/	NIH (National Institutes of Health) http://nih.gov	For downloading and sharing biomedical 3D print files, modeling tutorials, educational materials. Special collections: prosthetics, neuroscience, heart library, molecule of the month. Files are free.
Sketchfab http://sketchfab.com		300,000 models
Smithsonian X3D http://3d.si.edu/browser	Smithsonian Institution http://si.edu	Files created to serve as use cases to highlight collections and explore the Smithsonian's digital collection. Files are free.
Thingiverse http://www.thingiverse.com/	MakerBot http://www.makerbot.com	Over 100,000 models; best-known model repository.
YouMagine http://www.youmagine.com/	Ultimaker https://ultimaker.com	Over 10,000 models. Building a community for sharing and collaboration.
Yeggi http://www.yeggi.com		300,000 models. Search engine. Mouse-over shows repository and cost (if any). Ad driven.
http://nasa3d.arc.nasa.gov; http://3dr.aldnet.gov/	Additional U.S. federal agencies	Free, public domain. Vary in number or models and 3D printable models available.

Other Recommended 3D Model Resources

3Dagogo, affiliated with AstroPrint, requires that all 3D models be tested for printability before listing on the site. This repository will be attractive for libraries in which staff have not yet learned to critique models for printability.

3D Warehouse is an online repository for models created using SketchUp. 3D models are heavily weighted toward architectural designs, and each model page contains an embedded 3D viewer plus model analytics. Many of the models are not readily 3D printable but may serve as either inspiration or can be modified to print successfully.

Bld3r ("builder") is an online repository highlighting that they are open source and independent, unaffiliated with a printer manufacturer. All uploaded files are free to download, and weapons are forbidden. Bld3r provides 3D printing tutorials and allows models to be linked from another website, such as GitHub, rather than hosted on their site.

MakerHome and Hacktastic are blogs created by Laura Taalman. MakerHome is a blog that links to models created by Dr. Taalman—one a day from August 2103 to August 2014. MakerHome is now a closed archive, but Dr. Taalman continues to post models on Hacktastic. The models tend toward mathematical designs and practical objects. The blogs include hints and tools for modeling as well as specific descriptions of model and design failures.

NIH 3D Print Exchange is designed for downloading and sharing accurate biomedical 3D print files. Biomedical professionals can also benefit from using the discipline-relevant modeling tools and tutorials on the website, which also includes case studies that involve using 3D printed models as educational tools. Its special collections include prosthetics, neuroscience, a heart library, and a molecule of the month.

Sketchfab hosts a gallery to display 3D models with a subset available for download or to order a print using an external printing service.

Yeggi is a search aggregator for 3D printable models that searches several model repositories. Dating back to April 2013, Yeggi is supported through advertising, which is immediately apparent to users.

◎ Scanning Objects

The most efficient way to make a 3D print of an existing physical object is to first scan the item. Instead of using a flat paper (2D) scanner, 3D objects must be scanned with a 3D scanner to generate a likeness that can be reproduced in three dimensions. 3D scanners use laser sensors or photographs to make numerous measurements of an object's exterior, thereby creating a model of the object's surface. Once you have generated a scan, you will need to edit and clean up the model, fixing holes in the surface or removing extraneous parts, using 3D editing software, and then convert it into an STL file for 3D printing. This can be a very time-intensive process, but scanning makes it possible for users to make a duplicate of an existing object, such as a broken or irreplaceable part.

Tools used for scanning objects range in size, cost, and ease of use. The quality of a 3D model generated by scanning is dependent on several factors. Characteristics of the object and its surface and location, such as shininess or surrounding light and shadows, may not be controllable, and some objects may not deliver a good scan under any circumstance. Potentially controllable factors that may affect the quality of the generated model include the quality of the scanning device, the skill or patience of the person who is doing the scanning, and the willingness to clean up the data before exporting the result into a finished model. For more information on 3D scanning, see Salinas (2014) and 3D Systems Inc. (2015).

Several varieties of 3D scanners are available. They can be characterized by their portability, either desktop or mobile. Scanners can also be characterized by their method of modeling an object's surface, either using laser sensors or photographic images. Typically, desktop scanners tend to use lasers and mobile scanners tend to use photos, but this distinction is not always valid.

Desktop scanners are stationary and may require a direct connection to a computer with sufficient memory and processing power. An item must be brought to a desktop scanner to be scanned, so its use is restricted to portable items (see figure 3.2 for an ex-

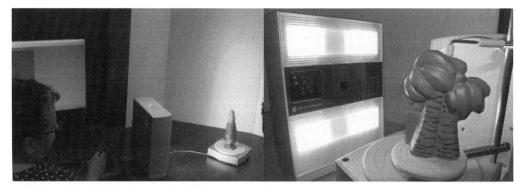

Figure 3.2. NextEngine Scanner, a Stationary Desktop Scanner. *Photos by Joe Baca (left) and Matthew Daley (right)*

ample of a NextEngine laser scanner at the University of Florida). If you supply a desktop scanner in your library, you must identify a location for its placement, perhaps dedicate a computer for accessing its software, and provide some level of staff or patron training in its use.

Mobile scanners come in several varieties and provide solutions for taking a scanner to an object. For large and unmovable items, mobile scanning is the only way (besides using a camera, repeatedly) to gather the digital data needed to construct a model. Mobile scanners are a viable solution for libraries that cannot afford or provide space for a larger and pricier desktop scanner, but that wish to provide some sort of scanning options for users. They can be circulated for use within or beyond the library.

If a 3D scanner is not available, a 3D scan of an object can be created through the use of just a tablet or smartphone. Autodesk's free 123D Catch app (http://www.123dapp.com/catch) works with still photos taken using a smartphone running the app or uploaded from a digital camera. Users take twenty to seventy photos of an object from all angles, preferably by adding newspaper or sticky notes to serve as reference points near the object. Then they upload those photos into a 123D Catch, process and clean up if desired, and then the app returns a 3D model. Similarly, the Rendor app (https://www.replicalabs.com/rendor) processes an uploaded video and then returns a 3D model.

Table 3.3 outlines some of the 3D desktop and mobile scanners that are commonly used in libraries. It also identifies some apps that can substitute for scanning devices.

Designing Your Own Model

Patrons and staff who are not quite ready to tackle designing their own model from scratch may begin by customizing some designs that have already been created. For example, in the Thingiverse repository (http://www.thingiverse.com), a subset of models may be altered using the Customizer feature, providing a very user-friendly way to alter or add words, names or initials, fonts, and sizes or number of moving parts of selected objects.

To 3D print an object that exists only in your imagination, you may create a model using 3D design software. There are varying levels of software programs for creating models, ranging in audience from beginner to expert. People with no related experience can use beginner-level programs successfully to create simple models. These programs also tend to be web based with free versions and advanced options available to paid users. Intermediate programs are less intuitive and have a longer learning curve than the beginner programs; however, they may be appealing to users willing to invest time because of additional func-

Table 3.3. Scanners and Apps

NAME	PRICE	NOTES
DESKTOP SCANNERS		
MakerBot digitizer http://store.makerbot.com/digitizer	$799	Scan objects 8" × 8", includes digitizer software, resolution 0.5 mm
NextEngine http://www.nextengine.com/	$2995 (base, requires software and accessories); educational discount	Resolution 0.1 mm, no preset size limit
Matter and Form https://matterandform.net/scanner	$599.99	Desktop but portable, scan objects 7" × 9.8", resolution 0.43mm, includes software
MOBILE SCANNERS		
Structure Sensor http://www.structure.io/	$379 or $499 with launch bundle	Mobile scanner attachment for iPads
iSense http://cubify.com/products/isense	$499	Mobile scanner attachment for iPads
Skanect http://skanect.occipital.com/	Free (but limited) version; $129 for single-user annual license	Software that turns a user's Structure Sensor, Microsoft Kinect, or Asus Xtion camera into a mobile 3D scanner
SCANNING APPS		
Autodesk 123D Catch http://www.123dapp.com/catch	Free mobile and web app	Upload photos, process as desired, and get a 3D model
Rendor https://www.replicalabs.com/rendor	In beta testing stage (Summer 2015)	Record a video, then upload through the Rendor app and get a 3D model

CASE STUDY: GOPHER TORTOISES VIA 123D CATCH

Gopher tortoises are an endangered species in Florida and are particularly susceptible to predators when juvenile due to their soft shell. Gabriel Kamener, a University of Florida undergraduate biology student, conducted a research study in summer 2014 to study these predation risks to young gopher tortoises. His plan was to place tortoise models in a study area and videotape interactions with predators, but Gabriel first had to make an inexpensive model using a single specimen preserved in alcohol. He couldn't use a desktop 3D scanner because the specimen couldn't be taken out of the alcohol for very long without degrading.

Gabriel used 123D Catch to 3D scan a tortoise specimen and printed the models at the UF Marston Science Library. He used the 3D model to create a cast, then duplicated the model using Plasticine since he wished to capture bite marks and the plastic was too hard for animals to bite. Gabriel then painted and scented the models, placing them in artificial burrows at the Archbold Biological Station in Highlands County, Florida. He successfully recorded predator occurrence and attack rates at each point using motion-sensor cameras. The primary predators were mammalian mesopredators, a majority of which were raccoons. View Gabriel's research video at https://vimeo.com/99487794.

tionality and features. Software programs that target professional users, such as engineers, tend to require local computer installation and licenses for individual or enterprise (institution-wide) use. Licensed versions may have an educational discount option.

Digital artists and engineers tend to have different design and functional needs for the objects they create; for example, artists may wish to develop smooth curves while engineers may require precision in size and fit. There are thus two major methods to designing a 3D digital model: one method is parameter based for precision, and the other uses sculpting techniques for creation of organic models. Parametric modeling software, such as Autodesk's Inventor and SolidWorks, allows a user to start with predetermined 3D shapes or a digital sketch that is extruded into a 3D shape. By resizing, rotating, and grouping shapes together, a designer can create 3D objects with exactly the dimensions required. Digital design software, such as Blender or Maya, allows more organic modeling capability along with texturing. 3D models are created through a process similar to a sculptor's working with a block of clay. Different digital tools are used to stretch and texture the model, allowing the development of natural models such as human faces and vegetation.

Two other modeling methods deserve special note since they will appeal to distinct subsets of patrons. OpenSCAD is an open source modeling software that contains a programming-style user interface where shapes are input using equations, and then modeling code is compiled to render the 3D object. Patrons who are comfortable with programming and are introduced to the wealth of online tutorials (e.g., MakerBot Industries LLC, 2015) may find OpenSCAD to be easily learned. For patrons (especially youth) who are familiar with building in environments such as Minecraft, the online program 3DSlash will appeal due to its interface that allows building 3D models by removing and adding blocks of varying sizes.

Table 3.4 lists popular software for creating 3D models. New software and updated versions appear continually, but the first software program a library should consider supporting and teaching is Autodesk's Tinkercad. Tinkercad is web based and freely available, requiring users to log in using an Autodesk account. Users under thirteen years old will need a parent's authorization to finish the registration process. Building 3D models in Tinkercad is very intuitive, since it involves combining and subtracting blocks of different shapes and sizes. Once the model is finished, Tinkercad allows exporting to an STL model that can then be 3D printed. Another advantage is that users can upload existing STL models and customize them in Tinkercad, such as add text or other shapes. Tinkercad is the ideal introductory CAD program to teach beginners of all ages and experienced modelers who would like a simpler program to quickly modify or create objects.

The concept of scale can be challenging for new users to keep in mind. The size of an object on the screen may correlate poorly with the dimensions of the finished printed object until the designer grows accustomed to using the rulers and other scaling tools present in the software. U.S. designers should note that most of the programs use the metric system, which measures in millimeters and centimeters, rather than the imperial system, which measures in inches. Another note of warning is that the units that export with the file are not always the ones assumed by the 3D printer software. If a 3D model is unexpectedly tiny, check to see whether the model was exported in inches instead of millimeters. Adjusting to scale may seem counterintuitive, but the important points for staff and patrons to keep in mind are (1) to assume that the scale is wrong if an object's size is far removed from expectations, and (2) to switch the scale to see if it helps, knowing the location of the all-important Undo button is if the switch isn't successful.

Table 3.4. Popular Software for Creating 3D Models

SOFTWARE	PRICING	FEATURES	TUTORIALS
BEGINNER			
3DSlash http://www.3dslash.net	Free web app, can register for more features, local app, premium version coming	If registered, can drag and drop your photos or import STL files to edit	https://www.3dslash.net/learn_tutorials.php
3DTin http://www.3dtin.com	Free, uses Google authentication		Click on Help (the ? icon) for a list of videos
Tinkercad http://tinkercad.com/	Free, requires account, must be age 13+	Recommended introductory software	https://tinkercad.com/video
INTERMEDIATE			
123D Design http://www.123dapp.com/design	Part of Autodesk 123D, free suite from Autodesk. http://123dapp.com		http://www.123dapp.com/design
SketchUp http://www.sketchup.com/	Make version is free, Pro version requires license, educational discounts available		http://www.sketchup.com/learn/
FOR DIGITAL ARTISTS			
Blender http://www.blender.org/	Free, open source	Good for rendering, sculpting, and animation	http://www.blender.org/support/tutorials/
Maya http://www.autodesk.com/products/maya/overview	Desktop subscriptions; from Autodesk	Supports animation, special effects, rendering, and shading	http://area.autodesk.com/mayalearningpath
FOR ENGINEERS			
Inventor http://www.autodesk.com/products/autodesk-inventor-family/overview	Free trials, desktop subscriptions, free for qualifying educational institutions	Intended for mechanical design	http://knowledge.autodesk.com/support/inventor-products/getting-started
Onshape http://onshape.com	Has free (student), professional, and enterprise levels	Full cloud (web browser) CAD system for collaborators	https://www.onshape.com/tutorials
OpenSCAD http://www.openscad.org/	Free, open source	Intended for engineers creating machine parts	http://www.openscad.org/documentation.html
Solidworks http://www.solidworks.com	Requires purchase but educational discounts	Core for engineers	http://www.solidworks.com/sw/resources/solidworks-tutorials.htm

BEST PRACTICES FOR SELECTING 3D MODELING SOFTWARE

- Consider any computer restrictions, such as the operating system, computer capability, and supported browsers. Some software is very resource intensive and your library computers may not have sufficient memory or processing capability.
- Select programs with license costs that fit your library budget. For expensive programs, check to see whether there is an educational license, or whether another part of your institution (especially for academic libraries) has already purchased a license.
- Try a variety of software to determine which interfaces and features will appeal to your staff and patrons.
- Determine what software your patrons are already using and consider supporting those programs or, at the least, developing familiarity with their capability and interface.
- For designing a specific 3D model, consider whether it is best created by grouping 3D shapes together or through "sculpting."

Modeling programs provide an option to export a completed model in a file format, typically STL or OBJ, which is readable by 3D printing software. Patrons whose model names include a file extension that is not understood by your 3D printer can be asked to check the repository for an appropriate version or to reopen their own model in modeling software and export it in a manageable format (see table 3.1 for common 3D file formats).

Many libraries have created miniversions of tutorials for creating models. These tutorials or videos are generally linked on library guides, and are sometimes embedded in 3D printing procedures or policy statements. Examples are highlighted throughout this book, and concentrated tutorial groupings are listed in chapter 11.

Key Points

Library staff will need to make referrals to modeling repositories and to recommend appropriate modeling or scanning software for patrons to use. Library staff are not required to become expert or even rudimentary modelers, but they will be in a position to give better service if they understand the fundamentals of modeling. The basic principles outlined in this chapter include:

- Design skills are not necessary; users can download free existing 3D models.
- When selecting a model from an online library, examine user-submitted photos to determine printability.
- 3D scanning options include desktop and mobile tools as well as cameras and free apps.
- Modeling programs range from free and simple to highly complicated and specialized, so an option exists for every potential user, budget, and purpose.

- Install free modeling software on your public computers and purchase additional software based upon the needs of your patrons.

Not every 3D model can be printed successfully. In the next chapter, you will learn how to examine a model and estimate its likelihood of successful printing.

References

3D Systems Inc. 2015. "3D Scanners: A Guide to 3D Scanning Technology." 3D Systems Inc. Accessed August 10. http://www.rapidform.com/3d-scanners/.

Bourke, Paul, 1999. "STL Format." Data Formats. October. http://paulbourke.net/dataformats/stl/.

———. 2015a. "Object Files (.obj)." Data Formats. Accessed July 27. http://paulbourke.net/data formats/obj/.

———. 2015b. "PLY—Polygon File Format." Data Formats. Accessed July 27. http://paulbourke .net/dataformats/ply/.

Grimm, Todd. 2004. *User's Guide to Rapid Prototyping*. Dearborn, MI: Society of Manufacturing Engineers.

Holdren, John. 2014. *Improving the Management of and Access to Scientific Collections*. Memorandum for the Heads of Executive Departments and Agencies. Washington, DC: Executive Office of the President, Office of Science and Technology Policy. March 20. https://www.whitehouse .gov/sites/default/files/microsites/ostp/ostp_memo_scientific_collections_march_2014.pdf.

MakerBot Industries LLC. 2015. "OpenSCAD Tutorials." MakerBot Support. MakerBot Industries LLC. Accessed August 10. http://www.makerbot.com/tutorials/openscad-tutorials.

McCune, Doug. 2014. "Using shp2stl to Convert Maps to 3D Models." *Code, Art, and Maps, Oh My!* (blog). December 30. http://dougmccune.com/blog/2014/12/30/using-shp2stl-to-convert -maps-to-3d-models/.

Peels, Joris. 2015. "Youmagine 3D Printing License 3DPL. Draft Version Number 8.1." Youmagine 3DPL. February 17, updated March 3. https://medium.com/@jorispeels/youmagine -3dpl-c11fce097ae.

Salinas, Richard. 2014. *3D Printing with RepRap Cookbook*. Birmingham, UK: Packt.

Singh, Parminder. 2015. *OpenGL ES 3.0 Cookbook*. Birmingham, UK: Packt.

Web3D Consortium. 2015. "X3D & VRML, The Most Widely Used 3D Formats." Open Standards for Real-Time 3D Communication. Web 3D Consortium. Accessed July 27. http:// www.web3d.org/x3d-vrml-most-widely-used-3d-formats.

What Makes a Good Printable Model?

ALTHOUGH IT MAY SEEM THAT ANY 3D MODEL should be printable, in practice not all models are suitable for 3D printing. Printing can fail for a number of reasons, including bad model design, poor printer configuration for a model, or hardware limitations of a particular 3D printer. This chapter will explore the best practices for designing and assessing models to ensure maximum printability.

Unsuitable Models

Just because a model is designed in 3D and can be saved as an STL file, does not mean that it is suitable for FFF-type printers, or compatible with all printers. Users must realize that there are some designs that will be impossible to print with a particular printer, due to size or resolution or basic limitations of FFF technology, no matter the user's skill at modeling. Figure 4.1 shows a model of a mathematical hypercube structure, with the left panel showing the 3D model and the right panel showing the print-preview image. An examination of this preview strongly indicates that this model is not suitable for a FFF-type printer because of the amount and placement of support material necessary to print. The central features are also at the same scale as the support material so the object and support material will merge, making it impossible to remove only the support material from the model. Another type of 3D printer, such as one with dissolvable support material or powder based, would be a better choice to use for printing such a model.

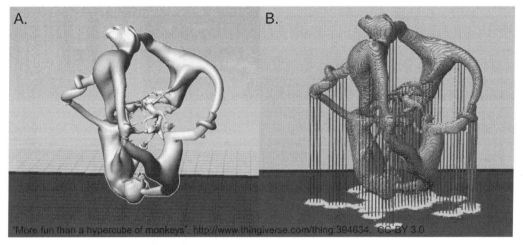

Figure 4.1. Hypercube of Monkeys

Even if a model is printable on a given 3D printer, the model may have errors or poor design that lead to print failure. Once a model problem is recognized, solutions include redesigning the model, trying a different 3D printer, and using model repair software. With experience, staff will learn how to identify these issues, ideally before the printing attempt, and will quickly learn the appropriate solutions to suggest to patrons. The next section discusses design principles that can help avoid print failure.

Design Principles

Good (or printable) models exhibit some basic characteristics:

- Walls are "watertight."
- Features are to scale and sufficiently thick.
- Model is customized as needed before submission.
- If necessary, model has been run through model checking or repair software.
- Model accommodates printer limitations.

These guidelines run the gamut from being absolutely necessary for successful printing to being recommended for an aesthetically pleasing or structurally sound model.

Watertight

A model's surface, which can be thought of as the model's skin, must be entirely continuous without any breaks in its surface. This requirement is termed watertight or manifold. Patrons may find this idea confusing and might ask if that precludes windows or a model such as a doughnut since those have holes. The distinction is that the surface of a doughnut has no breaks in the surface. A clean way of conceptualizing this is that if the surface has no openings such that water can penetrate the interior of the model, then it is watertight. A doughnut and a house with windows both include holes, but water cannot enter the interiors if the exterior structure is sound.

Figure 4.2 illustrates a model of a kidney that has gaps in its surface mesh and thus is not watertight. The printer software revealed a problem with the model by shading

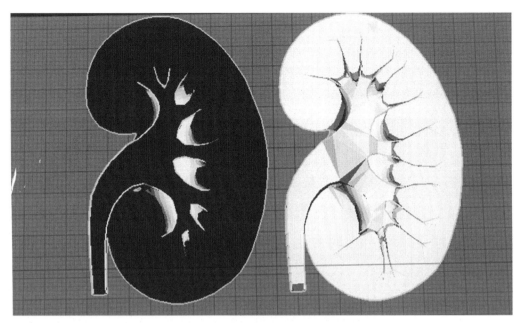

Figure 4.2. Model with Hole in Mesh

erroneous regions in gray. Processing software will vary in how it displays problematic models. However, if the model display appears unlike normal models, staff should encourage the patron to examine the model closely and use a model repair software (see below). If a model with a surface hole is sent to the 3D printer, the model either will not print at all, will skip portions of the model, or will print with a great deal of unnecessary support material.

Model Components

A model's walls and features must be suitably thick to be structurally sound. If a wall is too thin and thus only a tiny amount of filament is extruded, the resulting print will be wispy and too thin to attach to the surrounding walls. The minimum thickness that can be extruded will depend upon the printer's resolution capability because it limits the amount and accuracy of placed filament. However, that minimum amount may not be realistic to support a structure.

Figure 4.3 illustrates a model of a cell phone that appears to be error-free. The outer walls of the case, however, are designed to be about 1 mm thick, which can be easily determined since the fine lines in the grid of the printer software are 2 mm apart. Most 3D printers can print walls that are only 1 mm thick. What the patron will discover, postprinting, is that the walls of the resulting object are too thin and flimsy to withstand more than one attempt to insert a phone into the case.

Another crucial requirement for a model to successfully print is that the components of a model are connected. Features should not "hang" in the air; if so, the processing software will generate unnecessary support material to connect the floating areas.

Overhanging components that require support material should be minimized as much as possible if the user desires smooth surfaces or wishes to connect pieces to other parts.

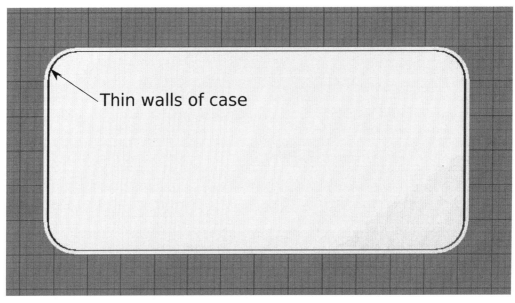

Figure 4.3. Too Thin: Phone Case

Features to Scale

A well-designed model should have features that are to scale compared with the actual object. For example, if designing a 3D model of a car, the wheels and steering wheel should be proportional to the entire car. The following case study illustrates the value of understanding scale on the part of modelers and model processors.

> In October 2013, the Shapiro Library at Southern New Hampshire University purchased its first 3D printer, a Makerbot Replicator 2X. Students, faculty, and staff were initially interested in the technology based on the novelty of 3D printing. A faculty member in the Marketing Department, who was teaching a Visual Merchandising class, came to library staff with questions and ideas about how 3D printing could be used to support her students' final projects. The final projects involved creating a model store front in a small box where miniature models of products, advertisements, and other store visuals were placed. Up until this point, students created miniatures for their final projects using clay or other materials or purchased miniature items as needed. This would be the first time that the Shapiro Library supported University curriculum with 3D printing.
>
> The Visual Merchandising students came to the library for a demonstration of 3D printing and some students selected 3D models, freely available online, to be printed including small mannequins, boots, snowboards, letters for signage, etc. Library staff had to consider the scale of objects that needed to be printed, which involved both trial and error efforts and mathematical calculations. Library staff realized through this project that there may be other items necessary to support 3D printing such as sandpaper, acetone, a heat gun, etc. For example, one student requested miniature snowboards which after printing, needed to be heated so that the ends could be bent upwards, resulting in a more realistic appearance. The students who requested 3D printing to support their Visual Merchandising final projects reported that the 3D miniatures supported and enhanced their projects. (Harris and Cooper, 2015)

The following textbox indicates some best practices for ensuring that models are set to the appropriate scale before being submitted for printing. These handy tips can be shared with staff and patrons.

BEST PRACTICES: ENSURING THAT MODELS ARE SET TO SCALE

Designing accurate models can be challenging when working abstractly within modeling software, but there are several techniques that your staff or patrons can use to ensure that a model is to scale:

1. Obtain measurements of real objects and then scale to the desired model size.
2. Choose model dimension units (millimeters or inches) that have physical meaning to you.
3. Look critically at the final 3D model and consider if all components are proportioned relative to each other.
4. Examine model dimensions visually using a ruler.

Large Models

Each printer has a maximum build size that determines the largest object that can be printed. This size depends upon the size of the build plate and the maximum distance between the extruder and plate. Printer software will give a warning message when an imported model is larger than what the printer can handle.

Solutions are either to resize the model to fit the build volume or to section the model into smaller components that can be glued postprinting. A good design practice for subdividing a large model is to consider the model's geometry. Instead of subdividing a model into equal pieces, it is better to divide along natural breaks in the surface to better mask the glue lines. For example, if sectioning a model of a snowman, divide into its natural unevenly sized balls rather than into equal thirds. More hints on subdividing models, especially to increase the strength of the resulting print, are included in a brief tutorial, which can be found at http://3dprintingforbeginners.com/stop-3d-printing-everything-in-one-piece/.

Letters and Numbers: Best Practices

Many patrons wish to customize pieces by adding names, dates, and other letter-based features to their models. Adding these features seems like the simplest form of model customization, but designing them to print successfully is quite a challenge. For best results, try the following practices:

- Lowered characters should be cut at least 2 mm deep.
- Raised characters should be heightened at least 1–2 mm above the surface of the object.
- Set characters to print on the horizontal plane whenever possible. Otherwise, support material will likely be needed and then will have to be cleaned out.
- Curved surfaces will need either to have the text wrapped around the surface or to have each character placed separately.

Consider printing an example for your service desk and workshops that is similar to the text calibration tool created by the authors in figure 4.4 (downloadable at http://www.thingiverse.com/thing:939461). The negative numbers in the top row correspond to the

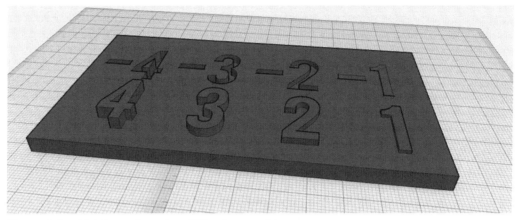

Figure 4.4. Thickness Samples of Numbers and Letters

millimeters deep that each number is cut. The positive numbers in the bottom row correspond to the millimeters high that each number is raised. Your example can be shown to patrons who inquire about the depth of letters and numbers in their own models.

◎ Developing Good Modeling Skills

Modeling Guidance

Chapter 3 described several types of modeling software and listed tutorials for their use. In addition to instructions in using the modeling software itself, several tutorials have been developed to assist in checking the resulting model for printability. 3D printer manufacturers and companies or libraries that offer 3D printing services have a vested interest in ensuring that models are in the best possible shape before submission to reduce wasted time, filament, and general unhappiness with the printer. Examples of general modeling tutorials are listed in table 4.1.

In addition to the detailed tutorials, several libraries note some basic principles of good design on their 3D printing guides. Examples of best practices are highlighted in the textbox.

Table 4.1. Modeling Tutorials

TUTORIAL	CREATED BY
3D Modeling for Beginners http://www.shapeways.com/tutorials/easy-3d-modeling-for-3d-printing-tutorial-for-beginners	Shapeways
How Do I Make a Solid Model? http://wiki.mcneel.com/rhino/faqclosedsolids	Rhino
How to Fix and Repair Your 3D Files http://www.shapeways.com/tutorials/how_to_use_meshlab_and_netfabb	Shapeways

◎ Case Study: Designing a Phone Case

Some objects raise several modeling issues. The case study illustrates an example in which the initial design resulted in an object that could not be printed successfully using PLA filament. Several redesign options were suggested to the patron.

Goal: Redesign to not have small raised surfaces on both top and bottom of object.

Challenge: Cell phone cases consist of a flat cover and curved edges that extend downward to hold the phone. If the cover is flat or if it has shapes cut out, such as letters or a silhouette, a phone case may be printed successfully on the flat side. However, if the designer wishes to include a raised design on the flat side of the case, the case will require a great deal of supporting material to print. Figure 4.5 illustrates the difficulty with printing out a case with a raised design using PLA filament. In both of the shown orientations, the model will require support material inside the case. Options for redesign include:

Solution 1: Remove any raised designs from the flat cover side.

Solution 2: Print any raised designs separately from the case, and glue them on after printing.

Solution 3: Use a 3D printer with dissolvable filament. It will still generate a great deal of support material, but the material can be removed cleanly without leaving rough edges.

BEST PRACTICES FOR CREATING SUCCESSFUL MODELS

- Save each piece as a separate STL file. Objects can be batched on a plate but cannot be separated if part of a single file.
- Export all models as STL or whatever format is accepted by your printer software.
- Save (or "save as") backups frequently while designing a model.
- Model as one piece or several—just be sure they connect and are not disjointed.
- Design with a solid base for your object.
- Pay close attention to scale and units of measurement.
- Check for and delete any 2D elements or floating pieces on your design so they are not saved as part of the model.
- Check the feature size. "In general, long thin features are difficult to print. Use this rule of thumb: for every 1mm of length a feature should be .5mm thick. So, a 5mm long robot's antenna would need to be about 2.5mm in diameter to survive once printed" (Northbrook [IL] Public Library, 2015).
- "The mesh or surface of the 3D model must be watertight and a solid. More technically, all faces of the object must construct one or more closed volume entities. When the faces are not fully closed, they produce gaps or holes in the model and those holes and gaps will keep the model from printing correctly" (Boise State University, 2015).

These best practices are taken from the websites and guides of several libraries (Boise State University, 2015; J. Willard Marriott Library, 2015; Northbrook [IL] Public Library, 2015; University of Arizona Libraries, 2015). Many of the practices are included on multiple guides.

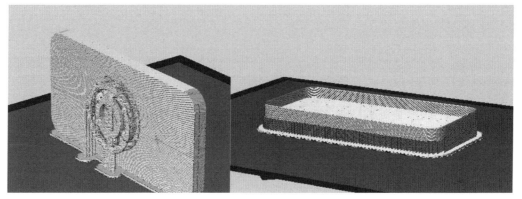

Figure 4.5. Bad Design: Phone Case Needing Supports on All Sides

Using Model Checking or Repair Software

The best way to ensure a high-quality, printable model is to have it checked by the true experts, that is, through computer software rather than the human eye. Encourage your staff and patrons to run their models through repair software at any time when the modeling process is not going smoothly, and especially when a model is completed. As noted in subsequent chapters, many libraries require that patrons run their models through a check/repair service before submission to save time, filament, money, and anguish on everyone's part.

When patrons submit a problematic model, they can either redesign to fix the error or use software to repair the model. There are many options for users, ranging from simple to complex (see table 4.2 for a list of repair software), but the first one users should try is Microsoft's Model Repair Service. It is cloud based and only requires a user to upload an STL file and then download the repaired file. Some model processing software, such as Simplify3D, includes some functionality for checking the model and repairing some or all of the surface mesh problems. If your processing software includes a checking or repair option, your staff should use it routinely.

Printing Setup

The majority of printer software contains default settings that will be sufficient for printing the average 3D model. However, for models that are complicated or will be used under demanding conditions (like a quadcopter body), there are several common printer settings that can be adjusted to ensure the best possible print. These settings include infill percentage, number of shells, inclusion of a raft or brim, plate placement, and whether to use support material.

The location of these settings will vary depending upon software and may be located in a custom print menu. There are also other advanced settings that should only be adjusted if a printer is having difficulty printing any models and are best left to highly experienced staff.

Consider making sample pieces that illustrate these concepts. They will assist your staff in explaining the settings and options to patrons. New users will learn from the visual aids, and patrons who are acquainted with the settings of other 3D printers will appreciate the differences in options among printer brands.

Table 4.2. Model Checking and Repair Software

SOFTWARE	VERSIONS
MeshLab http://meshlab.sourceforge.net/	Free, open source
Meshmixer http://www.123dapp.com/meshmixer	Free download
Microsoft Model Repair Service https://modelrepair.azurewebsites.net/	Free cloud service; requires Microsoft account
MiniMagics http://software.materialise.com/download-minimagics	Free download; requires registration
netfabb http://www.netfabb.com/	Private, Professional/Enterprise, Cloud, and other paid levels

Infill

Printed 3D models are normally not solid but rather filled with a pattern of varying percent density. Potential **infill** patterns range from honeycomb to grids, but not all printer software will offer the same pattern options. The density percentage will affect strength and rigidity of a model as it increases; however, it also adds to the mass of the model and the time to print. Users should be encouraged to stay with the default infill settings (depends on resolution, typically 10–30 percent) unless the model will be under stress and will require additional strength. See figure 4.6 for an example of how infill percentages vary from 0 to 100 percent. When you make similar samples of infill rates generated by your printer to show to your patrons, remember to stop the print job before the final top layers are finished.

Best practice: Use the default unless the model will be under stress and will require additional strength.

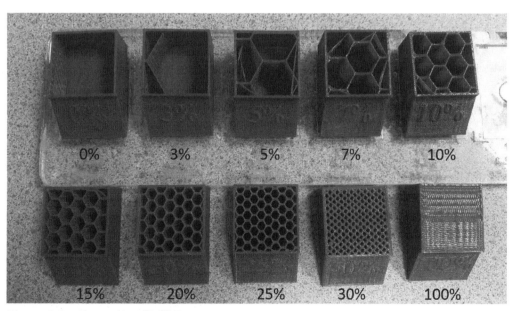

Figure 4.6. Examples of Infill Percentages

Shells

The **shell** of a model is the surface surrounding the infill. The number of layers that compose the shell determines its thickness and strength. Typically, the default or minimum number of layers in a shell is two. Additional shells will add strength and rigidity to the model but will require additional filament. Figure 4.7 shows a model of a box with two, eight, sixteen, and thirty-two shells. As the number of shells increases, the internal space filled with the honeycomb infill decreases.

Best practice: Use the default number of shells unless the model will be under stress and will require additional strength.

Rafts, Skirts, and Brims

Most printer software programs offer the option to lay down a **skirt**, **raft**, or a **brim** before printing the model. The **skirt** is a thin outline of the model's footprint, deposited on the plate before beginning the actual printing. The advantage of laying down a skirt is to clear the nozzle and ensure that a steady stream of filament is ready for printing. Some

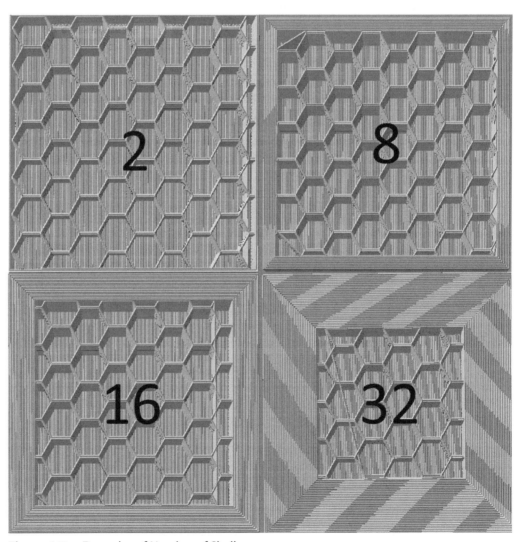

Figure 4.7. Examples of Number of Shells

software, such as MakerBot Desktop, causes the printer to lay down a line of filament instead of a skirt by default before a print.

A **raft** is a thin layer of filament, slightly larger than the model's footprint, deposited before printing the actual object. The raft provides support and stability to the object while it is printing, helping the object to stay securely on the build plate. It also facilitates removal of the finished object from the build plate, and can assist in keeping bottom corners and edges flat, especially when the build plate isn't flat or properly leveled.

A **brim** is a ridge added to the base of the model for several layers. It adds stability to a model, especially if the base is small and needs reinforcing. The brim, like the raft, is designed to break off from the model after removal from the build plate.

Figure 4.8 shows a 3D model of a teapot (http://www.thingiverse.com/thing:460243) sliced using Slic3r in the Repetier-Host software. The top panel shows the model sliced

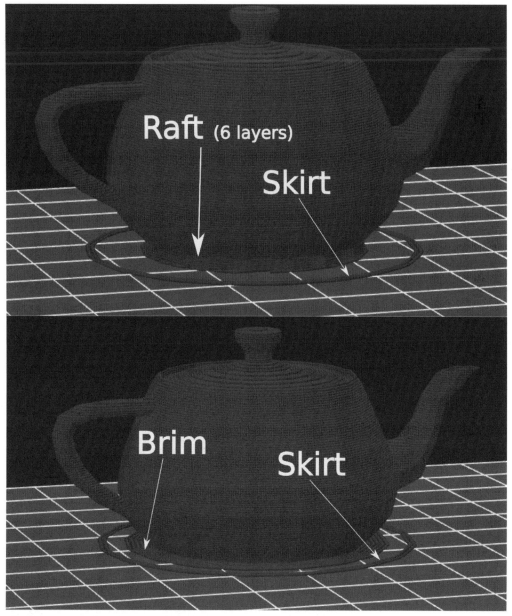

Figure 4.8. Rafts, Skirts, and Brims

with a six-layer raft (the model sits on top of the raft) and a surrounding skirt. The lower panel shows the same model now surrounded by a one-layer brim that extends 4 mm away from the teapot.

For further details and illustrations of these options, see Simplify3D's tutorial (2015). Many staff and patrons view the raft or brim as adding unnecessary time and cost to the job, and will wish to eliminate it during the model processing stage. However, if the lack of these support features causes a job failure requiring a do-over, both time and material are wasted.

Best practice: Always use a skirt if the processing software doesn't lay down a line by default. Rafts and brims should be encouraged to assist in adhesion to the print bed and to stabilize the print.

Support Material

3D printers can't print models in the air, and thus must use support material for features that overhang or have a large gap (or bridge). Not all 3D models that have bridges or overhangs require support material, and you may be surprised to discover what models can be successfully printed without support. The maximum bridging distance and overhang angle depends upon extrusion rate, temperature, and the printer's travel speed. These distances and angles are unique to each printer and must be determined through test prints. If you wish to explore the limitations of your printer, search Thingiverse.com for models designed for printing tests or conduct an experiment similar to 3DGenius's bridging test (Caplin, 2013).

Patrons may not want to use support material for several reasons:

1. The supports will be placed in a difficult-to-reach spot, making removal impossible.
2. Support material will leave a jagged surface and require sanding to remove. For example, support material inside screw threads requires a great deal of cleaning.
3. The amount of support material makes the print too costly or time consuming.

Library staff should ensure that patrons are aware of how much support material will be added to a printed model before printing, ideally showing them a print preview so they are prepared for the amount of postprocessing necessary. Patrons should also be asked if any sides of the model should remain smooth (such as the threads of a screw that ideally should be printed pointing upward, or whether a receptacle should be smoothest on the inside or the outside). It is best to include support material by default if the model contains overhangs or bridges to ensure the best odds for a successful print.

Best practice: When in doubt, process the model with supports and examine the resulting preview.

Model Placement on Build Plate

The placement and orientation of a model on the build plate is an important step for ensuring printability, ideally minimizing support material and reducing the likelihood of warping. Warping occurs when the edges of a model curl upward and is influenced by filament type, bed adhesion, and cooling rate. See chapter 10 for more discussion on how to minimize warping.

BEST PRACTICES: ARRANGING MODELS FOR SUCCESSFUL PRINTING

Staff will quickly learn to spot model features that are likely to lead to unsuccessful prints. Suggestions for best positioning of the model on the build plate include:

- Put objects near the center of the build plate.
- Cluster objects of the same height together on the build plate. If batching small pieces, bundle short pieces (with fewer layers) and taller pieces (with many layers).
- Group multiple objects close to each other.
- Place flattest side on the plate if possible.
- Lay object upright rather than horizontally if the object has small features on both top and bottom.
- Minimize the time that finished objects are left to cool on the plate.

Ideally, models should be centrally located, placed away from the plate edge, and with the flattest side resting on the plate. If a model lacks a flat surface, rotate the model to maximize contact with the plate. After rotating the model, make sure to check that the model is resting on the plate and not floating in the air. Some software will have a Drop to Build Plate button that should be used frequently. Be sure to inspect the model in the software from all directions, including below the plate, to check that it is resting on the plate and is not overlapping with other models.

Bundling several models together on a build plate does not decrease the total print time but does save handling time if all pieces can be successfully printed together. Avoid bundling complicated models with smaller and simpler models since, if the complicated model fails, all models may be lost.

🌀 Key Points

This chapter discussed several ways to improve the likelihood that models will print successfully. Options include following good modeling principles, using model repair software, and selecting appropriate printer settings for each model. Below is a list of best practices to keep in mind:

- Skillful modeling leads to successful printing.
- Whether designed by a patron or found in a repository, the benefits of running a model through model checking or repair software cannot be overstated.
- Patrons and staff will quickly develop basic skills at assessing whether a model looks printable.
- Don't assume that all models have been optimized for best printing. Look or ask.
- Careful placement of models on the printer build plate and use of rafts or brims to stabilize unsteady models can prevent print failures. Consult with the patron about preferred orientation of pieces, and check several arrangements on the build plate to minimize supports and maximize stability during printing.

- Print complicated models with several pieces as separate print jobs, rather than bundling. Then one unsuccessful piece will not hold up the other pieces.
- Adjust the percent of infill or number of shells to optimize for strength and stability versus cost and time to print.

Now that you understand the skill sets needed for modeling and assessing printable models, you are ready to begin planning and shopping for your 3D service. The next chapter will present the wide range of associated costs to assist you in planning a budget.

References

Boise State University. 2015. "3D Printing Request Form" and "3D Printing: Do's and Don'ts of Design." *3D Printing—LibGuides at Boise State University*. Boise State University. Accessed August 10. http://guides.boisestate.edu/3Dprinter.

Caplin, Steve. 2013. "A Bridge Too Far." 3D Genius: The Home of 3D Printing. http://www.3dgeni.us/a-bridge-too-far/.

Harris, Jennifer, and Chris Cooper. 2015. E-mail message to the authors. July 9.

J. Willard Marriott Library. 2015. "Basic Instructions 1"; "Tips And Tricks 1"; and "Tips And Tricks 2." *3D Printing Services—ULibraries Subject Guides at University of Utah*. University of Utah. Accessed August 10. http://campusguides.lib.utah.edu/3dprinting.

Northbrook [IL] Public Library. 2015. "Creating Your Own Design." Northbrook Public Library. Accessed August 10. http://www.northbrook.info/3d-printing/create.

Simplify3D. 2015. "Rafts, Skirts and Brims!" Tutorials. Simplify 3D. Accessed August 10. http://www.simplify3d.com/support/tutorials/rafts-skirts-and-brims/.

University of Arizona Libraries. 2015. "Prepare Your 3D File." Arizona Board of Regents for the University of Arizona. Accessed August 10. http://www.library.arizona.edu/services/print/3D/file.

CHAPTER 5

Estimating Expenses and Assessing Your Service

IN THIS CHAPTER

▷ Deciding how to charge for prints

▷ Estimating initial expenses of a 3D printing service

▷ Calculating a minimum budget

▷ Estimating recurring expenses

▷ Assessing your service and its continuing costs

BUDGETING FOR A 3D PRINTING SERVICE is a challenging task because each library's funding source, staffing patterns, user needs, and location options are unique. Rather than providing prices for highly variable components, this chapter will focus on outlining the variables that you should examine and include when calculating your likely costs.

Charging for Prints

If patrons pay for printing, a steady source of funding for ongoing service can be ensured. Consider modeling your charging structure (if used) on that used for existing charged services such as photocopiers and printers. A 3D printing service is parallel in many ways to a photocopy service, for which replacement supplies must be purchased as the service is used. See chapter 7 on policies for further discussion of the "to charge or not to charge?" decision.

A significant factor in determining the charge is whether the printer is paid for, or whether the service is expected to pay for the printer and perhaps to support additional staffing as well as continuing material costs. Libraries that have chosen to charge for 3D printing have developed a wide range of methods for charging for the use of a 3D printer. Charging models are based on material use, time, per job, or several of these elements, and some libraries establish a minimum charge. A selected range of charging options is presented in table 5.1. The data were gathered from library websites and from

Table 5.1. Selected Charging Models

SELECTED NO-CHARGE LIBRARIES		
LIBRARY	**NO CHARGE**	**NOTES**
Public Libraries: • Darien (CT) Library (resident children under 12) • Westport (CT) Library (residents) School Libraries: • Lake Forest Academy Academic Libraries: • University of Alabama • Boise State University • Kent State University • University of Memphis • University of Tennessee, Chattanooga	No charge	
McGoogan Library of Medicine	No charge	As supplies last
San Diego Public Library	No charge	Donations accepted

SELECTED LIBRARIES THAT REQUIRE PATRONS TO BUY OR BRING FILAMENT		
LIBRARY	**FILAMENT CHARGE**	**NOTES**
University of Michigan	BYO ($49/cartridge)	Also full-service option
North Carolina State University, D.H. Hill Makerspace (patron managed)	BYO or buy filament ($13.25 for .5 kg spool)	Also full-service option

SELECTED LIBRARIES THAT CHARGE BY MATERIAL USED		
LIBRARY	**FILAMENT CHARGE**	**NOTES**
• Cleveland Public Library • Riverdale Collegiate Institute Library	$.05/gram	
• Illinois Institute of Technology • Indiana University—Purdue University at Indianapolis (IUPUI) • Oregon State University • Public Library of Cincinnati and Hamilton County • University of Wisconsin, Stevens Point	$.10/gram	
University of Florida (August 2015–)	$.15/gram	$3 minimum charge
• Brigham Young University • University of Maryland • Miami University • SUNY Oswego	$.20/gram	
Darien (CT) Library	$.10 or $.25/gram, depends on printer	$2 minimum charge
Middletown (PA) Free Library	$.10/gram regular or $.20/gram special filament	
North Carolina State University, Hunt Library (staff managed)	$.35/gram or $10/cubic inch, depends on printer	$5 minimum charge
University of Tennessee, Knoxville	$.15/gram or $4.75/cubic inch, depends on printer	

SELECTED LIBRARIES THAT CHARGE BY MATERIAL USED		
LIBRARY	FILAMENT CHARGE	NOTES
University of Miami	$.25/liter (resin)	$5 flat fee/job
Fayetteville (NY) Free Library	$.05/gram $6/cubic inch + $3/build	For MakerBot Replicator For Mojo
University of Nevada, Reno	$8.45/cubic inch $2.61/ml of powder + $0.22 per ml of binder $5/cubic inch of powder + (if desired) $5/cubic inch of ColorBond	For Stratasys (ABS) printer For Z Corp printer plaster

SELECTED LIBRARIES THAT CHARGE BY JOB LENGTH		
LIBRARY	TIME CHARGE	NOTES
• Allen County (IL) • Southern Illinois University, Edwardsville	$1/hour	
North Dakota State University	$3/hour + $.05/minute afterward	$3 minimum charge
Westport (CT) Library	$10/hour for training; $10/hour for printing	Fee for visitors
Colorado State University	Tiered for one-time use, weekly, season/semester, visitor $10–$700	

SELECTED LIBRARIES THAT CHARGE BY BOTH MATERIAL AND TIME		
LIBRARY	MATERIAL AND TIME CHARGE	NOTES
University of Florida (5/2014–7/2015)	$.06/gram $.02/minute	$3 minimum. Changed (see above) after additional printer models acquired
University of Utah	$.04/gram $.50/hour	$3 setup fee

contributions to the Librarymakerspace-L discussion list. The table reflects the challenges of establishing costs when more than one type of printer is available. It also reveals that pricing is not correlated with library type.

Estimating Initial Expenses

The expenses to be incurred by each library in setting up a 3D printing service are highly varied because each library faces unique circumstances in funding, staffing, and space preparation. Table 5.2 outlines the elements to consider when planning a budget. Further details are presented below the table or in other chapters of this book as indicated.

Table 5.2. Budgeting Worksheet for Initial or One-Time Purchase Items

CATEGORY	NOTES	COST PER ITEM	YOUR COSTS
Space planning: • Room climate control • Printer climate control • Electrical power • Filament storage • Shelves, racks, cabinets • Security for printers and patrons	See chapter 2.		
Furniture: • Table for printer • Table for workspace	If you don't have suitable sturdy tables, you will need to purchase.		
3D printer	See table 5.3.	See table 5.3; typical range $1200–$3000.	
Computer and accessories: • Computer with graphics capability for processing jobs • Large or dual monitors • SD cards, flash drives, or other tools for transferring models and processed files		$1000–$2000 if new and dedicated	
Supporting software	E.g., Simplify3D.		
Filament or resin	See table 5.4.		
Common tools and supplies: • Scissors • Scotch tape • Flashlight • Sticky notes • Labels • Paper towels and cleaning wipes • Paper clips (unbent) • Straight pins and needles • Rulers marked with both inches and millimeters • Plastic bags: sandwich and gallon sized	Potentially raid or fund from general library supplies.		
Tools and supplies unique to 3D printing: • Acetone • Adhesive/glue/epoxy suitable for your filament • Build plate tape/glue/hair spray • Digital scale • Heat tool (intended for crafts) • Insulated glove • Pliers • Spatulas/scrapers: small, narrow edge, long handled, not too flexible	Likely unique to 3D printing service, and candidates for funding from the 3D operation.		

CATEGORY	NOTES	COST PER ITEM	YOUR COSTS
Staffing: • Handling general queries • Assisting patrons or managing the printers • Instructing: one-on-one and workshops	See chapter 8.		
Training: • Staff led • Paid instructors	See chapter 9.		
Outreach and events: • Marketing • Workshops • Additional	See chapter 11.		

◎ Space Preparation

The printer has to go somewhere, and it will include an entourage of accessories. Preparing the space may involve several initial expenses. Space preparation expenses may include adjustments to the printer location area to ensure visibility for staff and for patrons, safety to protect patrons from hot and moving parts, security to protect the printers and the print jobs from accidental bumping, and climate control for smooth and accurate printing. Enhancements to consider include enclosures, locks, or windows.

Your staff or patrons will appreciate a work area next to the printer. Since 3D printers come with an entourage of accessories, plan for storage space for filament, tools, and supplies. Shelves, cabinets, or rods will help control filament. Printer and library vendors offer carts that can house the printer, some filament, and some accessories. If you have SLA-type printers, you will need space for model cleanup and waste resin disposal, ideally with a washable surface and easily cleaned floors.

Furniture may be required for consultation or collaboration activities. Power supplies may need to be added, upgraded, or relocated. Some of these accommodations are pricey. Further details on space needs are outlined in chapter 2.

◎ Printer Costs

A wide range of printer types are available, and they can cost from $350 to six or seven figures. Chapter 2 discusses printer features in greater detail, since the selection process can rival the complexity of choosing a new vehicle. For a more detailed list of the basic features of these popular printers, consult table 2.3. Table 5.3 summarizes a list of 3D printer manufacturers with machines suitable for and popular in libraries, along with their pricing effective in the summer of 2015.

Table 5.3. FFF Manufacturers and Printers Popular in Libraries

COMPANY	URL	MODELS	PRICE
Afinia	http://www.afinia.com	H480 H800	$1,299 $1,899
Cubify	http://www.cubify.com	Cube CubePro	$999 $2,799
Flashforge	http://www.flashforge-usa.com	Creator Pro	$1,349
Formlabs	http://formlabs.com	Form 1+	$3,299
LulzBot	http://www.lulzbot.com	TAZ Mini TAZ 5	$1,350 $2,200
MakerBot	http://www.makerbot.com	Mini Replicator Z18	$1,375 $2,899 $6,499
MakerGear	http://www.makergear.com	MakerGear M2	$1,775
PolyPrinter	http://www.polyprinter.com	PolyPrinter 229 PolyPrinter 508	$2,395 $4,495
Printrbot	http://printrbot.com	Playtime Simple Metal Printrbot Plus	$399 $599 $1,199
Ultimaker	http://ultimaker.com	Ultimaker 2 Go Ultimaker 2 Ultimaker 2 Extended	€1,195 €1,895 €2,495
Stratasys	http://www.stratasys.com	Mojo UPrint SE UPrint SE Plus	contact for quotes

Computer and Software Costs

In addition to purchasing a printer, you will need a computer with sufficient graphics capabilities to process jobs. For example, the Simplify3D software, which can process models for a wide variety of printers, must be loaded on an "Intel Pentium 4 or higher processor, 2GB or more of RAM. Windows XP or greater, Mac OS X 10.6 or greater, Ubuntu Linux 12.10 or greater. OpenGL 2.0 capable system" (Simplify3D, 2015). You may choose to use or purchase a large monitor or dual monitors to better view the 3D models with your patrons. The software needed to process the jobs will come with the printer or will be free to download. But you may wish to purchase additional software to manage the printing operation as outlined in chapter 8.

Filament

Your printer warranty may specify that you need to purchase proprietary filament from a specific vendor for the life of the warranty. If so, you may have no other purchase options for a while. When and if you are not bound to sole-source filament, you may pursue several additional vendors that may offer quality filament at lower costs, more colors, and more filament types.

For nonproprietary filament, consult table 5.4 for a list of major filament suppliers and their costs for approximately one kilogram of filament, the weight of a typical spool. The price ranges reflect sale and special color options and specialty materials (such as wood or metallic components or glow-in-the-dark properties) rather than a basic price difference between filament types. Keep in mind that many of these vendors will offer a discount or shipping charge break for bulk purchases or for educational customers. Batching your requests rather than purchasing one or a few spools at a time will result in considerable savings.

⌬ Tools and Supplies Needed

Many 3D printers will arrive with their own tool kit, but not all kits will contain all the tools needed to launch a serious service. Check the printer documentation for suggestions of supplemental tools, and match it against your library's existing stock. Helpful tools include spatulas to remove finished jobs from the build plate, preferably those that are very thin at the leading edge but not too flexible. Heat-resistant gloves are a must if your printer has a heated build plate, and are convenient for various maintenance tasks.

Depending on the type of FFF printer, various supplies can be used on the surface of the build plate to aid in adhesion of the printed model to the plate. One inexpensive option is blue painter's tape, which can either be purchased in wide strips or even in widths specific to the build plate. Other possibilities include Kapton tape, a polyimide film backed with adhesive, or reusable surfaces such as BuildTak. It is extremely important to ensure that the layer is placed smoothly onto the build plate, without bumps or air bubbles. Glue sticks are recommended for glass build plates as well. If using ABS, models also adhere well to heated glass sprayed with hairspray (such as Aqua Net) or a thin layer of a slurry made of dissolving ABS with acetone. The ideal situation is for a model to sit firmly on the plate, avoid warping during printing, and then be easily removed without damaging the model or plate. The best solution for model adhesion will vary by type of printer, type of material, and even the model shape and size, so don't hesitate to explore other options if you are having difficulty with models warping or popping off the build plate.

Table 5.4. Nonproprietary Filament Suppliers

COMPANY	URL	COST FOR APPROX. 1 KG OF FILAMENT
Amazon (many vendors, including SainSmart)	http://www.amazon.com	$11.99 and up
ColorFabb (Netherlands based)	http://colorfabb.com	€35.95 for PLA/PHA
Fargo3D	http://www.fargo3dprinting.com	$20–$49 for PLA
Filaments.ca (Canada based)	http://filaments.ca	$28.95–$69.95 for ABS, HIPS, PLA
MakerGeeks	http://www.makergeeks.com	$21.55–$56.95 for ABS, HIPS, PLA
MatterHackers	http://www.matterhackers.com	$29–$98 for ABS, PLA
ToyBuilder Labs	http://www.toybuilderlabs.com	$30–$40 for ABS, PLA

Tools and supplies to enhance the surface of print jobs can be restricted to staff use or can be lent to patrons if you choose. Manual or electric sanders, pliers, craft knives, glues, and paints or nail polishes will prove helpful, but you will want to decide whether patrons must meet these needs on their own. Providing some finishing options for staff use will encourage buy-in for the service.

Digital scales are a highly useful accessory, especially if you have a FFF-type printer. In the early stages of calibrating a 3D printer, use it to verify the accuracy of the printer software's estimate of the amount of filament that jobs will require. Depending on the payment model, the model's actual weight might be used to calculate the price. Weighing finished objects will also help staff learn to eyeball what 5 grams or 170 grams look like. As spools of filament wind down, measure them (and don't forget to subtract the weight of the spool itself) and mark the remaining weight on the spool. Staff can then easily determine if an existing spool has enough filament to complete the next job. Most printers cannot recover a print job once the filament has run out, so it is especially crucial to verify that a spool has enough material before starting to print.

Additional supplies to complete your service include rulers marked with both inches and centimeters. Most 3D printers and modeling programs use centimeters and millimeters, and staff will appreciate having rulers stored near the processing computer to easily show patrons what a particular length in centimeters looks like. If you are not running a self-service operation, you may wish to invest in labels and self-sealing bags in various sizes (such as sandwich and gallon) to mark and store the finished jobs awaiting pickup. Calculators or spreadsheets will be appreciated by staff. Paper forms that contain patron and job information may help reduce confusion as you begin to accept several jobs in a day.

For an additional perspective on useful tools, consult the *Beginners Corner* blog (3dprintingforbeginners.com, 2015).

Minimum Budget

As this chapter indicates, estimating costs for a 3D printing service is challenging because each library has unique needs. If you need to purchase a computer or furnish a location, your start-up costs will increase significantly versus a library that can set the printer on an existing table and plug it in. For those who are fortunate enough to just plug and play, or who wish to launch a service as quickly and inexpensively as possible, table 5.5 illustrates that $1,500–$2,000 is sufficient for a start-up budget. Be prepared to restock filament and supplies quickly if you begin with small amounts.

Staff Costs

Initial staff time spent in preparing the proposal, learning to use the printers, crafting the policy and workflow, and planning and receiving training may be calculable with some reasonable accuracy. Recurring staff costs lie in providing direct patron assistance and in running or maintaining the printers. Additional time will be spent in developing and presenting marketing materials and workshops for patrons. Details on staff assignments are outlined in chapters 8 and 9.

Table 5.5. Bare Minimum Start-Up Budget

ITEM	DETAILS	APPROX. COST
Printer (midrange for libraries) + warranty	Table 5.3	$1,200–$1,500
Filament (6–9 spools)	Table 5.4	$200–$300
Unique tools and supplies (to supplement the samples that come with the printer)	Table 5.2	$100–$200
Total		**$1,500–$2,000**

Recurring Costs

Ideally, you have ordered sufficient supplies to see you through the first six to twelve months of service, but you will not be able to assess your precise usage until you begin printing. Table 5.6 provides a budget worksheet for noting the items you will have to accommodate in recurring budgets.

Assessing the Continuing Budget Needs

In addition to the recurring costs listed above, sustaining your 3D service requires additional costs, and you should account for them in the initial stages of budgetary planning. These include replacement or repair of equipment and supplies, new equipment for expansion of service, software upgrades, and continued marketing costs.

Replacement, Repair, and Additional Printers

Budget for repair and replacement costs, since you are likely to experience a need for them. Chapter 10 outlines some typical problems encountered when 3D printing, and even an extended warranty may not cover repair or replacement due to user failure.

Table 5.6. Budget Worksheet for Recurring Costs

CATEGORY	NOTES	COST PER ITEM	YOUR COSTS
Common supplies	See list in table 5.2. Refresh as needed.		
Filament	See table 5.4.		
Unique supplies • Compressed gas duster • Build plate tape, hairspray, or glue • Insulation tape	Varies by printer		
Replacement parts, e.g., nozzles	Varies by printer		
Staffing	If not absorbed by current staffing		
Outreach and events: • Marketing • Workshops • Additional			

The amount of time needed to print each piece is challenging to appreciate before you begin your printing operation, since it does not correlate at all with printing sheets of paper. Your 3D printing service may become so heavily used that you will quickly wish to add to your fleet.

A 3D printer is a toy as well as a tool, and therefore you are more likely to equate its attractive life span in a manner more similar to a personal vehicle rather than to a mundane paper printer or photocopier. You may consider it a lemon or a clunker, and you may become envious when you learn of the functionalities of other and newer types that have become available. Do not fight the urge to compare printers or to dream of more and better! As with any rapidly developing technology, better-performing models are available on what seems like a weekly basis. Pay attention to the features and functionality of newer printers that could have a strong, positive effect on your service, gather price quotes, and be ready to submit a purchase request when an opportunity becomes available.

Supporting Software and Other Equipment

The explosion of new 3D printers may or may not settle down soon. Regardless, printer owners will still need to keep aware of new software and equipment options to support the 3D printing operation. As mentioned in chapter 8, software to facilitate processing and sending jobs to multiple printers (such as Simplify3D) may become more developed and more desirable. Commercial products (such as Skyforge) are being developed and should improve in quality (if not in price), and libraries that have developed homegrown software to manage payment and file submissions may make their software available to others.

Marketing Costs

A 3D printing service may become one of those services that runs under its own power and popularity. But like all services, it may benefit from a boost in publicity from time to time. Ensure that you are including the service in your regular marketing channels, such as your website, social media, newsletters, and in-library signage and displays. Chapter 11 offers examples of marketing features presented at other libraries.

⑥ Evaluating Your Service

Usage Statistics

Maintain usage statistics for 3D printing that align with the data you gather for other library services. Summarize and report these statistics at the same interval used for other services: monthly, per semester, annually, and so forth.

Decide on what statistics you will find useful before you finalize your workflow procedures as outlined in chapter 8. Ensure that you are gathering needed data when you accept jobs, whether through forms or through staff input. Then, gathering or compiling the statistics at your specified interval should prove a straightforward task.

You will receive many queries or transactions about the service, and you are likely to want to capture this impact as well as the number of job submittals. Add a category for 3D printing queries to each of the service transaction types that you routinely record. Whether you record transactions by content type such as reference, directional, general, or by method received such as in person, phone, e-mail, or chat, add categories for 3D printing to each relevant block.

Table 5.7 summarizes some basic 3D printing service statistics that you may wish to record. If you don't routinely gather usage data but you see value in reporting on your 3D printing service, this table provides a starting point. Gather statistics on a monthly basis to reduce the size of the task.

User Feedback

In addition to recording usage statistics, you may also benefit from gathering qualitative feedback from both users and nonusers of the service.

Surveys

Surveys of your users may be of value at several points: before you purchase your printer, during the first year of service, and at regular intervals after the service is established. Surveys can be used to capture general interest, to gather explicit desires such as new filament colors preferred, and to capture assessments about the quality of your service and your staffing competencies.

Unsolicited Comments

File the comments gathered from social media or suggestion boxes. You never know when they will prove useful. Make sure your staff know what to do when they receive strong or precise verbal feedback about the service or a staff member. This feedback should be reported by following your existing procedures.

Table 5.7. Basic 3D Printing Service Statistics

BASIC STATISTICS	SOURCE OF DATA	NOTES AND EXAMPLES
Hours of use per printer	Printer	Each printer should have a counter. The trick is to check it at preferred intervals.
Number of print jobs	Your logs or forms	
Number of patrons requesting print jobs	Your logs or forms	
Number and percent of patrons per category	Your logs or forms	Age groups (youth vs. adult), academic disciplines (arts vs. engineering vs. social sciences), or however you categorize your patrons.
General inquiries about 3D printing	Your reference or general service statistics	

Key Points

No two libraries will have an identical budget for a 3D printing service, since fundamental supporting items such as a graphics-capable computer, an appropriately visible and secure location, and anticipated volume vary greatly. Any library service, no matter how popular, requires regular assessment and publicity to maintain its stable of users. This chapter focused on outlining items in addition to a printer that are needed to support a successful service. Keep in mind the following tips:

- A 3D printing service is attractive to many funding sources.
- Your costs will vary greatly depending on your setup needs and anticipated volume, but only $1,500 is needed to launch a bare-bones service.
- Staffing costs are real if unpredictable.
- Recurring costs are more easily measured once a volume of service is established.
- Gather and maintain usage statistics in line with your other record keeping.
- Keep marketing the service, even when the usage seems very high.
- If you don't have the time or skill set to keep up with the industry, rely on other librarians via discussion lists to continue learning about 3D printer technology and related library services.

The next chapter presents a variety of service models that you may consider adapting for your library. It discusses related service options and offers suggestions for prioritizing print jobs.

References

3dprintingforbeginners.com. 2015. "13 Tools to Help You Become a 3D Printing Pro." *Beginners Corner* (blog). March 11. http://3dprintingforbeginners.com/3d-printing-essential-toolkit/.
Simplify3D. 2015. "Checkout Questions." Support: FAQ. Simplify 3D. Accessed August 10. https://www.simplify3d.com/support/faq.

Developing Your 3D Printing Services

WHEN PLANNING FOR A NEW 3D SERVICE, many of the factors to consider will be unique to your particular library setting, patrons, and staffing. This chapter will cover different types of 3D printing service models and related services, and what depth or level of support you can provide for each. Your choices will depend on the amount of staffing you can afford to devote to each service, the needs of your users, the computer graphics capabilities of your staff and patron computers, and the expertise that exists or that can be developed among your staff.

Types of Service Models

One of the first questions to consider when developing the service and workflow is who will be responsible for processing files, starting printing, and managing the printer. Your service can be either fully staff managed, include partial participation by patrons, or be entirely open to patron usage. There are pros and cons for each style of service, and you will find examples of each across all types of libraries. Once you have decided upon the type of service your library will offer, chapter 8 offers in-depth guidance on how to structure your 3D workflow.

Staff-Managed Service

The first type of service model is one that is entirely managed by the library staff. Patrons bring models to the staff who process the files, start them on the printer, manage the printer, and remove the finished models to deliver to the patron. Within this scenario, there are two possible levels of staff mediation. In some libraries, patrons consult with the staff to decide upon the appropriate parameters for the print (such as resolution, size, color, and orientation). Alternatively, those choices may be left entirely up to the staff. Your library should also decide how patrons will bring the 3D files to the library staff, whether by USB drive or by connecting to e-mail or to a storage site.

A staff-managed service model has the advantage that patrons can readily fix any problem with a file as soon as it is noted by the processing staff. The patron can quickly go to a mobile device or to a library computer, resolve the issue, and return the file to the 3D staff for reprocessing. If staff process the files without patron interaction, you may wish to try an online service such as Skyforge (http://skyforge.co/) to manage the file uploading process, or set up a website form that includes either an e-mail address or upload capability to submit print jobs. It is wise to provide a method for patrons to approve the model print parameters since orientation on the build plate will affect the placement of support material, and the model size isn't always exported correctly from modeling software.

If the file is submitted online and patrons prepay for printing, you will need to respond with an amount and provide for online payment. If you charge patrons for printing, it is especially important for them to have the option to verify all aspects of the print job before they pay.

Once the model is processed using the printer software as described in chapter 8, staff then place the model into the queue and manage the printing. Staff then handle all printer tasks, including changing the material, preparing the build plate surface, and removing finished models from the printer. Patrons do not interact with the 3D printers or processing the models. Be sure to set a standard regarding the amount of postprocessing patrons can expect of staff, since some staff will be keen to separate the model from rafting and supports and others will want to leave it for the patron to handle. Allow patrons the opportunity to refuse postprocessing by staff since some models will be too fragile to easily remove extraneous material. Beware that some models may develop sharp edges if the support material does not fully detach and may pose a hazard to young patrons.

After the 3D model is finished printing, staff should place it somewhere secure and easily visible by patrons for pickup. Contact patrons once the models are finished for pickup, although you may discover that many patrons wait in the library for the printer to finish or anticipate the finishing time and return immediately for pickup. Determine an appropriate length of time that models will be stored by the library, awaiting pickup, to avoid building up a large collection of unwanted models, which is less likely to happen if patrons pay for printing.

Staff-Managed Example: Academic Library

The University of Florida Libraries' service model is fully staff managed where patrons bring STL files on a USB drive to the reference desks either at the Marston Science Library, the Health Science Center Library, or the Education Library during standard reference hours. The 3D service is open to all students, staff, faculty, and the community, with priority given to student assignments and then to the UF community. Patrons consult with trained staff to process the model, along with determining the orientation that maximizes contact with the build plate and minimizes support material. They also provide the correct size, filament color, and desired resolution. Patrons prepay, formerly at the rate of $.06/gram plus $.02/minute but changed to $.15/gram, using either a credit card or a UF purchasing card.

Library staff place the processed model file into the queue and manage the 3D printers. Staff print throughout the evening and occasionally after hours if necessary. Once the model finishes, it is placed in a bag with an identification slip and placed at the circulation desk for pickup. Each morning a staff member e-mails patrons whose models finished printing the previous day and updates the log to reflect the current print status (UF George A. Smathers Libraries, 2015). For more specifics about the workflow, consult chapter 8.

Staff-Managed Example: Public Library

The following points are excerpted from the "Lab Policy" at the San Diego Public Library (2015). These statements describe a process that appears very similar to that of the University of Florida.

- Patron must bring in .stl file format for objects to be printed.
- All files will be inspected and sliced by lab volunteer/staff prior to printing.
- Only Maker Lab staff/volunteers can operate the 3D printers.

◎ Patron-Managed Service

The second type of service model is one where patrons have full access to the 3D printers. They are responsible for preparing the models to print, handling the material, and managing the printer. To support self-service, the printer must be located in a space accessible to patrons, either in a supervised area or visible to staff to ensure safety and security. Allowing patrons to use the 3D printers unmediated poses logistical issues. Many print jobs experience problems, such as popping off the build plate or filament tangles, and these can be lengthy to solve. See chapter 10 for fuller descriptions of problems that affect all 3D printers. If you enact a reservation system to allow patrons fair usage, you will need to accommodate for unforeseen variability in model printing time that may result in one patron's printing time bleeding into another's scheduled appointment.

Another issue to address is how staff should handle situations where a patron monopolizes a 3D printer, interferes with another patron's printing, or if a model remains on the printer without the patron nearby to remove it. This is similar to challenges in a laundromat when clothing is removed prematurely from or left in a machine. Further, care must be taken to ensure that patrons do not "tinker" with the printer or software configuration. You will need to develop a policy regarding printer damage, whether through carelessness or intent.

If patrons are responsible for payment, they can either purchase the printer material outright or pay for only the amount consumed. Some printers will provide a measurement on material consumed, such as the Cube filament cartridges, or staff can weigh material before and after to calculate usage. If patrons provide their own filament, you must ensure that they only use material that is suitable for the printer, since the use of unapproved filament may damage the printer or void the warranty.

Libraries that offer patron-initiated service tend to require those patrons to have attended a library-sponsored workshop to learn about 3D printing in general and the library's procedures in particular. An additional technique to reduce failure during patron-initiated service is to require that each model has been run through a model repair software program to ensure its structural viability before submission.

Patron-Managed Example: Academic Library

Patrons can print on their own at the Morgan Library at Colorado State University after receiving training. The two-hour-long certification is offered twice a week or by appointment and teaches 3D printing, safety, and how to use the 3D printers. Once certified, users may print on their own while the library is open. Print jobs are restricted to students, faculty, staff, and affiliated guests for academic or curricular purposes (Colorado State University Libraries, 2014).

Patron-Managed Example: Public Library

At the Fayetteville (NY) Free Library, "The public often relies on the library staff for their initial training and introduction to some of the equipment in the Lab. . . . However, many people who use our space, after attending an initial certification class, are totally self-guided and self-led, and the community of makers using our space frequently learn from and teach one another in a way that goes broader and deeper than what our staff expertise would support" (Fayetteville [NY] Free Library, 2014b).

Hybrid and Multiple Services

A hybrid service workflow provides for some patron participation while retaining oversight of the printing process by staff. Staff may provide guidance and assistance with the processing of the 3D model to prepare for printing and allow the patron to handle the actual printing. Another option is to provide workshops and training for patrons to develop expertise with the 3D printers and allow only these trained patrons to have access to unmediated printing. This lessens the workload for staff and allows the enthusiastic and knowledgeable subset of a library's patronage to use the 3D printers by themselves.

A potential additional benefit of this approach is that these trained users may be willing to volunteer to print additional models for other patrons, further decreasing the workload for staff. Often, libraries will begin with either the patron- or staff-managed models and then merge into a hybrid model as the service evolves. Don't hesitate to switch or merge models if you discover your staffing or demand requires modification to sustain your 3D service.

Staff-Patron Hybrid Example: Westport (CT) Public Library's 3D Service

The public library in Westport, Connecticut, offers 3D printing to the public based upon a reservation system. Volunteers who are experienced with the 3D printers (known as "3D printer coaches") train patrons on how to use the printer and solve any problems that occur. After training, the patron is allowed to schedule a two-hour session to use the printer while a coach is available (Westport [CT] Library, 2015).

Multiple Options

Some libraries offer both staff-mediated and patron-managed options. For example, the University of Michigan offers both "self-service" and "full service," using different printers for each type of service (University of Michigan 3D Lab, 2015). Likewise, North Carolina State University offers full service at the Hunt Library and "do-it-yourself" service at the D.H. Hill Makerspace (NCSU Libraries, 2015). Providing more than one type of service works more smoothly if printers or areas are designated for each option.

⊚ Choosing Your Service Option

Obviously, there are many permutations of these service models. The best scenario for your library will depend on the size of your 3D printing service relative to your staff availability, type of printer, skill and interest of your patrons, and desired payment model.

Consider a staff-managed service if your library:

- Wants or needs to locate the printer in a space inaccessible to patrons.
- Needs patrons to prepay for models.
- Can't afford to replace the printer if mismanaged by patrons.
- Has staff willing and available to accept model submissions.
- Anticipates patrons submitting large and complex models.
- Plans to let the printers run while the library is closed.

Consider a patron-managed service if your library:

- Has a place to locate the 3D printer that is visible to staff and accessible to patrons.
- Can provide training or certification for patrons to ensure they know how to use the printer.
- Has a system to manage patron reservations.
- Does not plan to charge for models.
- Has a relatively user-friendly 3D printer such as a MakerBot Replicator, Cube, or Leapfrog Creatr.
- Lacks available staff to accept and manage 3D model submissions.

A hybrid service or multiple types of service are appropriate if your library:

- Has a mixture of user-friendly and more complicated or expensive 3D printers (such as a powder-based 3D printer).
- Serves a range of patrons, including children who require supervision with the printers.
- Has a 3D service with a demand that is higher than what staff alone can support.

Table 6.1. summarizes the pros and cons of offering each type of service.

Table 6.1. Summary of 3D Printing Service Types

TYPE	DESCRIPTION	PROS	CONS
Staff-Managed	Staff manage all aspects of the printing, including accepting and processing models.	• Removes safety concerns of patrons handling printers. • Printing is more efficient and less concern of printers being damaged by untrained users.	• Patrons don't gain hands-on skill with printers. • Heavy burden on staff time.
Patron-Managed	DIY: Patrons use the 3D printers by themselves.	• Lesser burden on staff time. • Patrons take greater responsibility for failures.	• Patrons may try to print bad or lengthy models. • Patrons may not follow instructions. • Complicates charging for models. • Printers may quickly become damaged and unusable. • Queueing challenges: Printer problems may push back appointments.
Hybrid	Patrons allowed to use printers with staff oversight.	• Permits patrons to work with equipment according to their comfort level. • Good option if library has multiple printers where some are suitable for patron usage and others need to be restricted to staff.	• Moderately heavy burden on staff time. • Queueing challenges.
Multiple	Offers more than one of the options above.	• Same as for individual options. • Satisfies more patrons by letting them choose their preferred option.	• Same as for individual options. • Likely to require multiple printers.

Expansion of 3D Service

A 3D service can be limited to simply providing availability to 3D printing, or it can be expanded to include additional expertise and service. Potential additions to your service include:

- Modeling software on selected library computers
- Teaching workshops about 3D printing, scanning, and modeling
- Model consultation with selected staff
- Scanning tools and software in the library or available for circulation

While it might seem tempting to restrict your service simply to 3D printing, you may discover that patron expectations (see chapter 10) will encourage you to broaden the service. At a minimum, your staff will need to know when and how to refer patrons to 3D modeling and scanning tools.

Many modeling software programs are available freely to patrons (see table 3.4 for list), but some libraries may wish to provide access to these programs on computers, ideally with large monitors and sufficient graphical support to render models quickly. Academic libraries can consult with their campus computing unit, which may already provide access to advanced modeling software for students. Many problems with 3D models are discovered when the submitted model file is being processed in preparation for printing. It is therefore convenient for patrons to be able to fix flawed models immediately in the library. If this is not possible, it will be important to provide links to introductory level cloud-based software tools, such as Tinkercad or 123D Design, which patrons can use to create a model within a browser window.

Regardless of whether your library opts to offer computers with access to modeling software, you may still want to have a staff member develop expertise with modeling to offer support to patrons. See a list of online tutorials in chapter 3 and library-created tutorials in chapter 11. Another way to build up expertise in the library is to seek patrons who are willing to share their knowledge and recruit them to teach workshops or offer advice one-on-one with other patrons seeking help. This expertise might be delivered in the form of general training sessions or through individual consultations to discuss particular model problems. For example, the 3D Studio at Northeastern University's Snell Library offers model consultation through drop-in hours Monday–Friday, 9:00 a.m.–5:00 p.m., and Saturday 1:00 p.m.–3:00p.m. Patrons can bring in an existing model or receive basic guidance on 3D printing from the studio's manager (Northeastern University Libraries, 2015).

With many software programs available, it is a challenge to be familiar with all the tools a patron might decide to use. However, it is worth surveying your patrons to discover what software the majority prefer to use, as a small number of programs may support a large fraction of users. If your library serves a broad range of patrons, consider choosing one or two options for beginners, one type to meet the precision needs of engineers, and one blend-and-smooth type for artists. At the University of Florida, the majority of 3D library staff have experience with Tinkercad and a few have basic familiarity with SolidWorks, the CAD program used by most undergraduate engineering students. For other types of 3D modeling software, it is useful to be aware of how to export and convert model files to STL or the format needed in your 3D printer processing software. See chapter 3 for more details about software and various 3D model file types.

Beyond software and model assistance, libraries may want to offer scanning tools to assist patrons with creating 3D models. There are several different methods for scanning in models as presented in detail in chapter 3. If scanning hardware is desired, you will need to either set up a dedicated area in the library for patrons to scan their objects, or circulate devices for patrons to borrow. For tabletop scanners, you should set up a partition or wall surrounding the device to prevent the light from disturbing patrons.

Image-based scanning techniques, such as 123D Catch, can be supported by either circulating tablets to allow users to take photos or by providing training and support for taking and processing the images. Mobile scanners can be attached to iPads, and you may wish to consider purchasing some of these tablet attachments for circulating to your patrons.

ⓖ Hours of Service

The decision of which hours to allow 3D printing will affect a range of other issues, including when models are accepted for printing. You may not want to allow printing after hours until you have sufficiently tested your equipment and feel certain that your printers will not malfunction while unsupervised. There are very few reports of 3D printers catching on fire (Antslake, 2014; Gonetotox, 2014), but you do not want to discover lesser evils such as a mess of "spaghetti" or hours lost to "air printing" after filament breakage when you return to the printer. Chapter 10 illustrates more disaster scenarios to prepare for when 3D printing.

If you choose to allow after-hours printing, you should initially be conservative with the type of models allowed to print unsupervised. Low-risk models for after-hours printing include those you have printed before, ones that are small in size, and ones that are directly downloaded from an online source with evidence of successful prints (such as user reviews on Thingiverse.com). At some point, you will need to decide whether to accommodate large models that will take longer than your printing hours to complete, or to restrict patrons to printing smaller models. Alternatives to unsupervised printing include either slicing the model into smaller components that can be glued or pegged together after printing, or decreasing the print time by lowering the resolution or decreasing the model size.

Another consideration is what time period you will accept 3D models for printing, assuming that staff process the models. Model consultation and processing can take a significant amount of time, with most jobs requiring at least ten to fifteen minutes for the processing and model payment. Some patrons will bring multiple models, and it isn't unusual at the University of Florida to spend an hour with a patron. Remember that models can also take a great deal of time to slice using the printer software, depending on resolution, complexity, and the speed of the computer.

Depending on your library's level of staffing, these modeling consultations can prove to be a lengthy burden on staff, especially if there are time periods where your desk is short staffed, such as in evening or weekend time periods. It is best to restrict model submissions to periods of maximum staffing, especially when staff who have expertise with the 3D printers are available. Figure 6.1 displays a sampling of hours during which five public and five academic libraries offer 3D printing services as of June 2015. The length

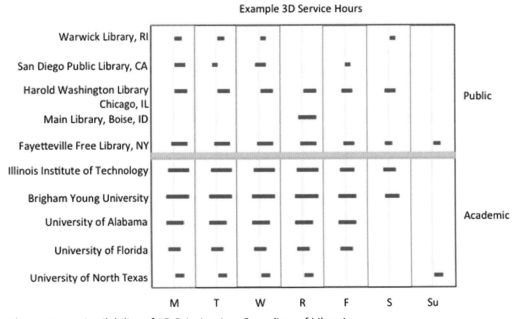

Figure 6.1. Availability of 3D Printing in a Sampling of Libraries

of a bar denotes range of hours either when 3D printers are available to patrons, or when staff are available to accept 3D models for printing. Hours are gathered from the Boise Public Library (Moeller, 2015), Brigham Young University (BYU, Harold B. Lee Library, 2015), the Chicago Public Library (2015), the Fayetteville [NY] Free Library (2014a), the Illinois Institute of Technology, Paul V. Galvin Library (2015), the San Diego Public Library (2015), the University of Alabama (Scalfani, 2015), the University of Florida (UF George A. Smathers Libraries, 2015), the University of North Texas Libraries (2015), and the Warwick Public Library (2015).

Case Study at the University of Florida

At the University of Florida, the science reference librarians accept 3D models for printing from 8:00 a.m. to 4:00 p.m., Monday through Friday. The reference desk closes at 5:00 p.m. but, due to the rush of patrons at the end of the shift and the time needed to process each order, the 3D hours were moved back to "last in line by 4:00 p.m." to allow enough time for processing before many staff need to leave. Night and weekend staff are permitted to accept jobs from patrons, but only as their time, attention, and staffing allow. Patrons can pick up models at the general service desk during business hours.

Figures 6.2 and 6.3 illustrate the distribution of 3D patron visits and questions throughout the week in the 2015 spring semester (January–April). Notice the peaks in job submissions and general 3D questions on Mondays and Fridays in figure 6.2. Although the 3D desk, collocated with reference, only officially accepts submissions from 9:00 a.m.—to 4:00 p.m., staff do accept them at other times in the day. Shown in figure 6.3, the majority of submissions are in the afternoon from 1:00 to 5:00 p.m. Notice the substantial peak at 4:00 p.m., which is after the submission period has closed. Staff report lines of patrons arriving near 4:00 p.m., waiting to submit models to 3D print. Some outliers indicate the time that staff processed jobs rather than the patron interaction time.

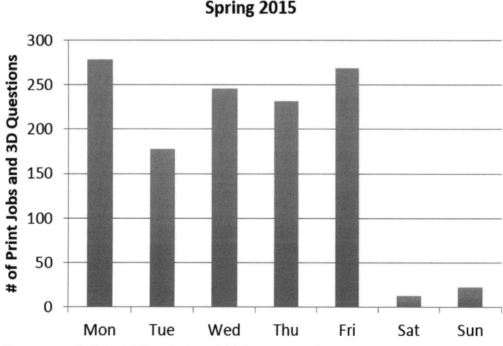

Figure 6.2. Daily Variability of 3D Model Submissions and Questions Received

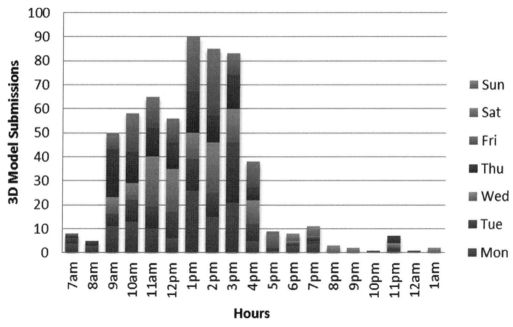

Figure 6.3. Hourly Model Submissions throughout the Week

At academic libraries, a further consideration is that demand will fluctuate over the course of a semester, with significant spikes in requests occurring in concurrence with class projects and at the end of the semester. All types of libraries can expect spikes before gift-oriented holidays. At the University of Florida, during the spring 2015 semester, peaks in model submission were seen around deadlines for major engineering classes and at Valentine's Day.

Prioritizing Print Jobs

At the beginning stages of a 3D service, you are likely to handle models in the order received. Once a queue begins to form of models waiting to print, you may choose to develop a system of prioritization. There are several factors to consider, and each library will have a unique set of preferences that determine a priority policy. One is to identify your users and decide if a subgroup will receive priority over others. An academic library may prioritize students over other staff or faculty. A public library may decide to give priority to local users or to teens.

You may also want to base priority on the intended use of the model. At the University of Florida, students with assignments have the highest priority followed by research projects. Other scenarios to consider include assigning a lower priority to patrons who submit multiple pieces or very lengthy print jobs. You may choose to squeeze short jobs in front of very long jobs, or group jobs based upon the requested filament color.

Best Practices for Successful Queue Management

- Group together models that use the same color or type of filament to save time swapping filament spools between print jobs.

FACTORS FOR DETERMINING PRIORITY

- Type of patron; e.g., primary vs. visitor
- Length of printing; e.g., short jobs to squeeze in vs. long jobs that might run past staffing time
- Patron's deadline
- Purpose of model; e.g., assignment, contest, gift, fun
- Quantity of models submitted; e.g., only print one copy of an item at a time
- Color of model, based on what color/type is currently in the printer
- Time dropped off or submitted; i.e., first come, first served
- Time the job is processed, which is not always correlated with time submitted. Some jobs require unusual or repeated processing.
- Reservations made vs. walk-ins

- If you allow printers to run throughout the night, finish small jobs first and then set the longer jobs to print after the library closes.
- Identify your highest-priority patrons and ensure they receive preference in the queue.
- Merge single jobs in with multiple-item jobs so all patrons are served equally.
- If a model continually fails, print another patron's model while you troubleshoot so that the queue does not become too backed up.
- Make sure to ask patrons if they have a deadline and place it in the queue accordingly.
- Advertise the lead time needed and the typical queue time so your patrons learn to plan ahead and to submit jobs with sufficient lead time.

Planning is required to prioritize the queue, but it will lead to more efficient printing when optimizing color/filament changes, staff availability to supervise the printer, and attempts to meet patron-requested deadlines. This policy should be clearly communicated to patrons so that they understand how their model will be handled in comparison to other submitted jobs. University of Florida patrons sign a user agreement stating they understand and agree to this policy, and staff verbally explain to each patron where their model fits into the existing queue. Chapter 7 includes more examples of policy statements that address prioritization. The textbox indicates factors to consider when choosing whether to prioritize some print jobs over others.

Key Points

When developing your 3D service operation, keep the following tips in mind:

- Each type of printer management—staff managed, patron managed, or hybrid—has its own set of challenges.
- Base any expanded/related services on the skill sets of your staff or reliable volunteers.

- Patrons' preferred hours for the 3D printing service may differ from other library services.
- Develop a clear policy for how to prioritize the model queue.

Now that you understand the various types of service models, the next chapter points out the public policy issues involved in running a 3D printing service. It also discusses considerations for creating your own policies, including examples from libraries already offering 3D services.

References

Antslake. 2014. "Yes, 3D Printers Can Go On Fire." *RepRap Forum: Safety and Best Practices.* January 14. http://forums.reprap.org/read.php?392,294850.

BYU, Harold B. Lee Library. 2015. "3D Printing Guide." *Subject Guides at Brigham Young University.* Brigham Young University. Accessed August 10. http://guides.lib.byu.edu/3Dprinting.

Chicago Public Library. 2015. "Harold Washington Library Center: Maker Lab." Harold Washington Library Center. Chicago Public Library. Accessed August 10. http://www.chipublib.org/maker-lab/.

Colorado State University Libraries. 2014. "3D Printing @ Morgan Library." Colorado State University Libraries. Accessed August 10. http://lib.colostate.edu/3dprinting.

Fayetteville [NY] Free Library. 2014a. "FFL Fab Lab." Fayetteville Free Library. http://www.fflib.org/make/fab-lab.

———. 2014b. "Makerspace FAQ: 21 Questions about FFL Makerspaces." Fayetteville Free Library. http://www.fflib.org/make/makerspace-faqs.

Gonetotox. 2014. "Fire Hazard." *RepRap Forum: Safety and Best Practices.* June 28. http://forums.reprap.org/read.php?392,375707.

Illinois Institute of Technology, Paul V. Galvin Library. 2015. "3D Printing." *Research Guides at Illinois Institute of Technology.* Paul V. Galvin Library. June 8. http://guides.library.iit.edu/3D-Printer.

Moeller, Kathy. 2015. "Library's 3-D Printers Popular to Check Out." *Idaho Statesman.* January 24. http://www.idahostatesman.com/2015/01/24/3607968/librarys-3-d-printers-popular.html.

NCSU Libraries. 2015. "3D Printing." North Carolina State University. Accessed August 10. http://www.lib.ncsu.edu/do/3d-printing.

Northeastern University Libraries. 2015. "3D Printing Studio." Northeastern University Libraries. Accessed August 10. http://library.northeastern.edu/services/3d-printing-studio.

San Diego Public Library. 2015. "Maker Lab." *Resource Guides.* San Diego Public Library. August 4. http://sandiego.communityguides.com/MakerLab.

Scalfani, Vincent. 2015. "The University of Alabama Libraries 3D Studio Standard Operating Procedures." Version 6.0, July. http://guides.lib.ua.edu/ld.php?content_id=5524864.

UF George A. Smathers Libraries. 2015. "3D Printing at the Libraries." *Guides @ UF at University of Florida.* University of Florida. July 13. http://guides.uflib.ufl.edu/3dprinter.

University of Michigan 3D Lab. 2015. "3D Printing." Digital Media Commons. University of Michigan. Accessed August 10. http://um3d.dc.umich.edu/?p=8520.

University of North Texas Libraries. 2015. "The Factory." University Libraries. Accessed August 10. http://www.library.unt.edu/services/factory.

Warwick Public Library. 2015. "3D Printing." Warwick Public Library. Accessed June 5. http://www.warwicklibrary.org/idea-studio/3d.

Westport [CT] Library. 2015. "3D Printers." Westport Library. Accessed August 10. http://westportlibrary.org/services/maker-space/3d-printers.

Policy Development

WHEN DEVELOPING YOUR LIBRARY'S 3D PRINTING POLICY, you do not need to start from scratch, since many libraries have created policies that contain text you may wish to borrow or adapt. Be sure to read several policies and read them all the way through. Some statements may make more sense in the context of that library's total workflow. Even if you do not wish to borrow from other policies, you will gain insight into the issues that you should resolve before you launch your service.

Legal Considerations

ALA and Public Policy

The American Library Association (ALA) has crafted two documents regarding public policy issues for 3D printing services in libraries. These reports by ALA serve as helpful background and context for developing local policies by discussing the implications for supplying 3D printing services in libraries.

The ALA tip sheet "Progress in the Making: An Introduction to 3D Printing and Public Policy" (Wapner, 2014) describes some exciting products that have been created by 3D printers in libraries. It then highlights some legal implications of intellectual property and liability issues, and briefly addresses concerns related to intellectual freedom and individual liberty regarding what types of objects should be permitted or restricted to print. The sheet concludes with a suggestion that, in the absence of law or jurisprudence related to 3D printing, libraries should begin by developing best practices that are consistent with the mission of the library.

The lengthier report "Progress in the Making: 3D Printing Policy Considerations through the Library Lens" (Wapner, 2015) explores more thoroughly the public policy issues such as educational applications and the economic implications of this one-off technology. Libraries are typically cheaper places to print than other makerspaces that may require membership fees. The disruptive influence of 3D printing on large-scale manufacturing is not yet known, but the impact on small businesses and individuals could benefit through well-balanced public policy. The report stresses the need to maintain values regarding intellectual freedom and free expression while safeguarding intellectual property and rights-protection issues such as copyright and infringement, along with a dash of product liability concern. This balance parallels that required for other equipment and services traditionally offered in libraries, such as photocopiers and word processors. A sidebar by Deborah Caldwell-Stone (Wapner, 2015: 8–9) recommends upholding these conceptual values while safeguarding user safety, equitable access, and protection from liability. The report offers cautionary points regarding the complexities of running a 3D printing service, and the responsibilities that libraries incur in instructing patrons not only in printing and modeling but also in the implications of the "social, economic, technological and public policy implications of this technology" (Wapner, 2015: 14). Extensive notes provide links to additional reading. An appendix prepared by Tomas A. Lipinski (Wapner, 2015: appendix) recommends that librarians post by their 3D printers a notice regarding copyright as is typically posted near photocopiers, such as the one he shares. He also provides a legal context for the notice.

At Your Institution

Consult other policies in effect at your library or institution that might cover legal topics. Any copyright warnings that you currently post near photocopiers, scanners, and other reproduction equipment should be repeated or adapted in your 3D printing policy and signage.

If your institution has developed a computer use policy that governs the type of items that are not permitted to display on your monitors (such as those that create a hostile or disruptive environment), or other "acceptable use" policies, these can and should be adapted to also govern the 3D models that you will accept and print. At the very least, ensure that your 3D printing policies are compatible with any existing user behavior policies in effect at your library.

Align with Other Services

After you have covered the legalities, turn your attention back to your library and examine its mission and the public services you currently offer. Your existing policies may include language that will help formulate your 3D printing policy. Some potential parallels include:

- Are all of your services available during every hour that your library is open?
- Do you offer some consultation services that are available only when those "experts" are around?

- Do any of your services require an appointment-only basis?
- Do you charge for special services or do you collect fines for overdue items?
- Do you provide a different level of assistance with (known) class assignments than with other inquiries?
- Do you have policies, written or unwritten, regarding when to hold patrons' hands and when to back off to let them do the work?

Issues Unique to 3D Printing

Some service issues are unique to 3D printing. You may not be in a position to flesh out all of these issues until you have received and tested your printer and established your workflow. Not all of these details need to be included in your policy. Many of them are best practices that are more appropriate to address in your procedures or workflow, especially if they are likely to evolve or to require judgment by staff. Including these issues in conversations about policies will help ensure that your policies are covering the critical aspects of your service.

Some libraries' policies include (or consist primarily of) procedural guidelines and workflow steps, and the rationale for doing so is understandable. In the first place, 3D printing policies and procedures are highly intertwined, along the lines of "We won't

CORE ISSUES TO RESOLVE

Issues unique to 3D printing that may drive your service policy and practices include resolution of the following elements:

- Types of service: printing, modeling, etc.
- Level of service: basic to advanced, to individuals or to groups/workshops
- Hours of support available for each level of service
- To charge or not to charge?
  - What charging structure will you use?
  - Payment mechanism
- Handling failures and do-overs
  - Automatic reprinting of obvious printer failures, vs. contacting the patron first
  - Refunds or vouchers issued?
  - Different if printer/staff failure vs. model/user failure?
- Priorities
  - For types of jobs—such as class assignments or meeting other deadlines
  - For users—open to anyone, or just your registered users?
- What not to print
  - Copyright restricted, license or patent violation, unlawful
  - Anything prohibited by an existing computer use or acceptable behavior policy

print unless you have followed certain steps." In addition, librarians may choose to not burden their patrons with two sets of instructions to follow.

Your policy should address any of the issues above that are likely to cause confusion on the part of your patrons. Some points, such as types of service or hours or costs, can be clarified with basic signage, handouts, or online guides. However, both staff and patrons will benefit if elements such as do-overs, priorities, and exclusions are included in the policy.

You may choose to ask patrons to sign or check off an acknowledgment of your policy at the time of their first submission or with each submission. This acknowledgment will reduce the complaints and tears that will result from the inevitable mechanical and human failures involved in the 3D printing operation.

Internal policies should not be included in the public policy. You may, however, wish to address some staff-related items in your staff training and procedures manuals. Will you permit staff to print their own items at no cost, and where will staff jobs fall into the queue? Are staff (some or all) permitted to make judgment calls regarding do-overs or inappropriate requests, or must all questionable issues be referred to the coordinator or senior staff?

◎ Learn from Other Libraries' Policies

Several libraries have posted some or all of their 3D printing policies on their websites, and selected examples are listed below. The statements reproduced here may have changed since this writing, but they continue to illustrate the variety of elements that some libraries have included. These representative policies illustrate the range of detail and highlight the issues considered of greatest concern to those libraries. Links to these libraries' policies are included in a list near the end of this chapter.

Allen County Public Library Policy

Excerpts from the Allen County (IN) Public Library (Spring 2015), where two branches offer a patron-operated makerspace that houses more than just 3D printers.

What are the hours of the Maker Lab?
Main Library Maker Lab hours are Monday–Thursday 1:00–9:00 and Saturday from 10–6:00. (Certified Adult users will be allowed into the Lab in the morning if the Lab is available.)
Georgetown Maker Lab hours are Monday–Thursday 4:00–8:30 and Saturday from 11–5:30.

Who can come to the Maker Lab?
Anyone can come to the Maker Lab; however, there are some things you should know before coming:

- Children under the age of 12 need to have a parent or legal guardian with them
- No open toed shoes please! Some of the materials and equipment we work with could hurt your toes if they were exposed; we like you to have as many digits when you leave as when you came in.
- No loose-fitting or dangling pieces of clothing. We don't want your favorite shirt getting caught in machinery.

- Tie back long hair. We wouldn't want your hair getting caught in machinery any more than your favorite shirt.

How can I become certified?

The Maker Labs at Allen County Public Library offer a wide variety of introductory courses. Once a patron takes one of these classes, they become certified in that piece of equipment or software. Certified adults can book time to use that piece of equipment independently, at any ACPL Maker Lab that provides it. Certified minors between the ages of 12 and 18 can also book time to use equipment independently, but only when a lab is staffed. If you are younger than 12, not only must the room be staffed, but a responsible adult must accompany you at all times.

Patrons can become certified in as many different technologies as they please, in fact we encourage it!

What is the cost?

All workshops are free, but advanced registration is needed, patrons will need an email address.

3D Printing is $1.00 for things printing in less than 60 min., $2.00 for things printing in less than 120 min., $3.00 for things printing in less than 3 hours, etc.

San Diego (CA) Public Library Policy

"Lab Policy" (Spring 2015) for the San Diego (CA) Public Library, which has four 3D printers:

- Patron can have 1 item per day to be printed by staff/volunteer, up to 2 hours a day.
- For print jobs longer than 2 hours, appointment is required 1 week in advance.
- 3D printers are available for use on a first come first served basis.
- Patron must bring in .stl file format for objects to be printed.
- All files will be inspected and sliced by lab volunteer/staff prior to printing.
- Only Maker Lab staff/volunteers can operate the 3D printers.
- Computers in lab are for scanning and slicing 3D objects only.
- Lab is volunteer run and is operated based on donations.

Westport (CT) Library Policy

Liability Waiver form (adult version) for the Westport (CT) Library's Makerspace Program in Spring 2015. The Westport Library is a public library with three 3D printers.

I, _____, am fully aware that participation in __Maker Space Program__ may result in risk of personal injury or harm.

I hereby agree to release and hold harmless the WESTPORT LIBRARY ASSO-CIATION, its officers, employees, volunteers, committees and boards, from and against any and all liability, loss, damages, claims, or actions (including costs and attorney fees) for bodily injury and/or property damage, to the extent permissible by law.

This indemnification and hold harmless agreement shall include indemnity against all costs (including without limitation, reasonable attorney's fees and court costs), expenses and liabilities incurred in or in connection with any such claim or proceeding brought thereon and in defense thereof.

I have read and understand this release, indemnification and hold harmless form. I voluntarily sign it and hereby give permission to the WESTPORT LIBRARY ASSO-

CIATION for emergency transportation and/or treatment in the event of illness or injury. I hereby accept responsibility for the payment of any emergency transportation and/or treatment. I further certify that I am in good physical condition, and have no medical or physical conditions that would restrict my participation in this activity or program.

Riverdale Collegiate Institute Policy

"Our Rules" (from Spring 2015) from the Riverdale Collegiate Institute (Toronto, ON) Library, which has one 3D printer in a grades 9–12 school:

1. Every .stl design file must be sent to Netfabb 3D repair to "fix" the .stl file for printing.
2. You import your fixed .stl file into Makerware. You would generally do this on the library MacBook (the library has the SD card for the printer).
3. In MakerWare you make print choices (such as size and scale, which extruder nozzle to select, whether your design needs a raft or supports to print properly and more). The 3D Club members can help with those choices.
4. From Makerware, you create your final .thing file and load it onto the Makerbot SD card.
5. Finally, you need to arrange a time you can print it (including when staff is available for the time needed).
6. YOU MUST BE THERE TO SUPERVISE THE PRINTING PROCESS. This is a great time to do homework (as long as one eye is on the print!)
7. The cost to you is the plastic, at 5¢/gram, so most items under 5cm³ will be well under $1. It's the TIME that costs!

Lake Forest Academy Policy

Below is the policy (from Spring 2015) from the Lake Forest Academy (IL), which is an independent grades 9–12 school.

File approval:
This 3D printing service is limited to currently enrolled LFA students, staff, and faculty. All submissions are subject to approval based on scheduling and availability. Files will be printed in the order that they are approved, not the order that they are submitted. Approval is pending the review of the print request by Mr. [A]. Academic projects have priority over all other requests. Please note that we are not able to print more than one project at a time. Multiple orders from one person will be staggered in our print queue. No objects designed as weapons will be printed.

Estimated Time Frame:
Please note that printing times vary based on the number of requests in the queue, complexity and totality of the design, and staff time and material availability. Please allow at least 3–4 days for printing when requesting an item. Many items will be faster, and some may take longer. A time estimation will be included in the project approval.

Quality:
Items printed on the Makerbot may have small surface defects such as bumps or holes. Please also note that while the Replicator 2x is very accurate, we do not guarantee any precise tolerances on fitting of multi-part objects.

Cost:
   3D printing services are available to all LFA students, faculty, and staff free of charge!

Size:
   The Makerbot can build objects with dimensions equal to or less than 5X5X5 in. However, our staff can help you create multi-part objects that can be put together to make a larger item.

University of Florida Libraries Policy

Three libraries at the University of Florida have 3D printers, with more on the way. The libraries' policy (Spring 2015) covers a completely staff-mediated service:

When submitting a 3D print job, you will be asked to sign a user agreement and to indicate your department and status at UF as well as a contact email. The agreement reads:

- I acknowledge that my item(s) adhere to copyright restrictions and to UF's computer use policies. I understand that the Libraries reserve the right to decline a print request that does not comply.
- I understand that I will not be reimbursed for jobs that failed on my part, but I will receive a voucher for jobs that fail on the library's part.
- I understand that the Libraries cannot guarantee a print time; however, staff may be able to provide an estimate based upon the print queue length.
- I understand that priority is given to UF students, faculty, and staff, and also to class assignments.
- I understand that items printed that are not picked up with 14 days will become the property of the Libraries. Items must be picked up by the individual who printed them.

University of Alabama's Policy

Excerpts from the University of Alabama's policy (Spring 2015) governing 3D printing only, for a patron-operated service. Additional printers are available on campus, and this policy only covers the printers in the Rodgers Library.

Users: The 3D Studio is available for use by all current University of Alabama faculty, staff, and students.

Operation and Safety Training: Participation in a Rodgers Library 3D Printer Training and Safety Workshop offered by a Rodgers Library staff member is required before using the 3D Printers independently. Training will be offered throughout the regular academic year. A schedule for the 3D Printer Training and Safety Workshop is posted at the Rodgers Library Circulation/Reserve desk. The 3D printer safety guide on page 9 must be completed and returned to the Rodgers Library circulation/reserve desk before experimenting with the 3D printers in the studio.

Availability, Scheduling and Starting Jobs: 3D print jobs can be started between 8 AM–6 PM on M–F. 3D printers can be left to run overnight (M–Th) only if Rodgers Library is open 24/5 (see our hours here: http://www.lib.ua.edu/hours). Note that Rodgers Library is not open 24/5 during the summer and other interim breaks. After attending the Rodgers Library 3D Training and Safety Workshop, new users will be added to an authorized 3D printer user list.

As an authorized 3D printer user, you may experiment with the 3D printers independently. However, please do not hesitate to ask questions and seek further assistance from us at any time. The 3D printer tool kit, manual, and log book are available at the front circulation/reserve desk for checkout to authorized 3D printer users.

Only authorized 3D printer users may schedule print jobs online. . . .

Responsibilities of Authorized Users: Authorized users are to follow all operating and safety training administered. It is expected that users will be considerate of others and maintain a safe working environment in the 3D Studio. Accurate note taking of print jobs and scanning should be recorded in the Rodgers Library 3D Studio log books. See log book forms . . . for required data logging.

Cost: 3D printing and scanning is completely free. Enjoy!

What Can and Can Not Be 3D Printed/3D Scanned: Users may 3D print or scan any object for research, education, or personal hobbies with the following exceptions: 1) No weapons or weapon accessories; 2) No illegal objects; and 3) No copyrighted objects or any other objects that violate patents and trademarks.

University of Maryland Library Policy

Excerpts from the University of Maryland's instructions (from Spring 2015) for one staff-mediated 3D printer:

- 3D printing is the process of creating physical objects from a digital model. The cost of printing is 20¢/gram. We will be happy to work with you to set up a file of the object you wish to print. If the job does not work because of machine or file error on our part, we will be happy to reimburse you or print another model. We are not however responsible for errors caused by other instances, such as corrupted files or improper file formats, unusual objects and other unforeseen issues.
- If you already have a file created, it must be in STL, or stereolithographic file format. Builds exceeding five hours duration will need to be assessed by TLC staff. You will be able to pick up your finished model at the desk the day after it is finished printing (varies depending on job length).
- Printing jobs are currently on a first come, first serve basis. Please let the TLC desk staff know if your job is a rush and we may be able to expedite it.
- Slight imperfections, such as the small indents and rough edges at the base of an object, are to be expected.

McGoogan Library Policy, University of Nebraska Medical Center

Policy from the McGoogan Library of Medicine at the University of Nebraska Medical Center (Spring 2015), which covers one 3D printer in a staff-mediated service:

1. The 3D printer is available for use by current students, faculty, and staff of the University of Nebraska Medical Center and Nebraska Medicine.
2. All use of the 3D Workstation must be reserved in advance (Click on the link in the box to the left).
3. Only designated staff will have hands-on access to the 3D printer.
4. Staff are acting as facilitators to provide access to the 3D workstation and printer. Staff do not necessarily have the skills to assist you with its operation especially during evening hours.

5. This service is designed for rapid prototyping, not production. No jobs printing large numbers of identical objects will be accepted.
6. All printed objects must be picked up at the AskUs Desk within 30 days of printing.
7. The 3D printer may be used for academic or business-related purposes only.
8. Objects printed on the library's 3D printer must be done so for lawful purposes. Users must abide by all applicable intellectual property rights and laws including copyright law (Title 17, U.S. Code) and patent law (Title 35, U.S. Code), UNMC policies, and library policies. McGoogan Library staff reserve the right to decline any print request for any reason.
9. No firearms or other weapons, obscene or inappropriate objects may be created using the 3D printer.
10. The McGoogan Library cannot guarantee model quality or stability, confidentiality of designs, or specific delivery times.

Examples of specific points from selected libraries' policies are listed in the next textbox. Some of these statements appeared in the fuller policy excerpts above, but they are repeated here to illustrate how different shades of language are used to convey similar ideas. These points reflect some of the unique aspects of 3D printing services that are less likely to occur in existing policy statements. The options under "We will/won't print" illustrate a range of general to detailed wording and examples of both positive and negative spins that can be used, and your choices of phrasing might be best influenced by your existing policies.

Following is a list of libraries with links (valid in Spring 2015) to the policies excerpted above and several additional examples. As is frequently the case, library policies begin to look derivative after the first few reads. Several libraries crafted their policies before the ALA reports were published, and future or modified library policies may begin to include more elements of evolving public policy. Keep in mind that some policies apply to a fuller makerspace as well as to just a 3D printing service, and the policies will reflect the types of printers and services offered at each location. Peruse several policies in their entirety to understand the context of each point. Look across library types, since policy similarities and differences do correlate with the type of library.

Allen County (IN) Public Library	http://www.acpl.lib.in.us/home/maker-labs
Brigham Young University (UT)	http://guides.lib.byu.edu/3Dprinting
Cleveland Public Library	http://www.cpl.org/EventsClasses/3DPrinting.aspx
Colorado State University	http://lib.colostate.edu/3dprinting
Darien (CT) Library	http://www.darienlibrary.org/services/3d-printers
Fayetteville (NY) Free Library	http://www.fflib.org/make/makerspace-faqs
Gloucester County (NJ) Library System	http://www.gcls.org/makerstudio/policies
Illinois Institute of Technology	http://guides.library.iit.edu/3DPrinter
Kent State University (OH)	http://libguides.library.kent.edu/3d
Lake Forest Academy (IL, grades 9–12)	http://lfanet.libguides.com/content.php?pid=560139

The following statements appear in various libraries' policies.
   We will/won't print:

- "The Library reserves the right to refuse any 3D print request."
- "The Library's 3D printer may be used only for lawful purposes."
- "Users may 3D print or scan any object for research, education, or personal hobbies with the following exceptions: 1) No weapons or weapon accessories; 2) No illegal objects; and 3) No copyrighted objects or any other objects that violates patents and trademarks."

Unfortunate incidents are possible:

- "Staff are acting as facilitators to provide access to the 3D workstation and printer. Staff do not necessarily have the skills to assist you with its operation especially during evening hours."
- "I understand that I will not be reimbursed for jobs that failed on my part, but I will receive a voucher for jobs that fail on the library's part."
- "The Library cannot guarantee model quality or stability, confidentiality of designs, or specific delivery times."
- "Every .stl design file must be sent to Netfabb 3D repair to 'fix' the .stl file for printing."

Users and priorities:

- "Use limited to . . ." / "Priority given to . . ."
- "No jobs printing large numbers of identical objects will be accepted."
- "Reservations required."

Safety and training (these statements are more appropriate for fuller makerspaces or patron-initiated printing services):

- "Children under the age of 12 need to have a parent or legal guardian with them."
- "No open-toed shoes; no dangling clothing; tie back long hair."
- "Participation in a . . . Training and Safety Workshop . . . is required before using the 3D Printers independently."
- [Patron] "must be there to supervise the printing process."

McGoogan Library of Medicine, University of Nebraska Medical Center	http://unmc.libguides.com/content.php?pid=653308
Miami University (OH)	http://www.lib.muohio.edu/computing/3d-printing
North Carolina State University	http://www.lib.ncsu.edu/spaces/makerspace
Oswego State University of New York	http://www.oswego.edu/library/services/makerbot_request.html
Riverdale Collegiate Institute (Toronto, ON) Library (grades 9–12)	https://goo.gl/w83h8W
San Diego (CA) Public Library	http://sandiego.communityguides.com/MakerLab
Southern Illinois University, Edwardsville	http://siue.libguides.com/content.php?pid=348515&sid=2851133
University of Alabama	http://guides.lib.ua.edu/makerspaces
University of Florida	http://guides.uflib.ufl.edu/3dprinter
University of Maryland	http://www.lib.umd.edu/tlc/tlc-tech-desk
University of Memphis (TN)	http://libguides.memphis.edu/3dlab
University of Michigan	http://um3d.dc.umich.edu/
University of Nevada, Reno	http://campusguides.unr.edu/content.php?pid=424521&sid=4337282
University of North Texas	http://www.library.unt.edu/services/factory
University of Tennessee, Knoxville	http://libguides.utk.edu/content.php?pid=601081
University of Utah	http://www.lib.utah.edu/services/knowledge-commons/3d-printing.php
University of Wisconsin, Stevens Point	http://libraryguides.uwsp.edu/3dprinting
Westport (CT) Library	http://westportlibrary.org/services/maker-space/3d-printers

Approval and Updating

Once your policy is drafted, run it by the library administration for approval and to catch any issues you may have missed. Your administrators may choose to share the draft policy with the institution's legal team for an additional layer of approval. As with any new technology, new or updated legislation is slow to catch up with practice, but identifying similarities with other services will increase the odds that your 3D printing policy will align with your institution and library's missions.

During the first year of operation, you are likely to question your policy on a quarterly basis. You will encounter transactions that cause you to question the wisdom of some points in your policy or that will simply encourage you to write a new point or adjust an existing point. If you add more printers or new types of printers, your policies should be reviewed for needed changes. Once established, examine and evolve your policy at least annually or as frequently as you examine other policies.

◎ Key Points

The term "policy" can refer both to a library's internal service policies and to greater public policy issues, so make sure all parties understand the meaning of the moment. Best practices for both aspects of policy include:

- The ALA's Office for Information Technology Policy (OITP) continues to develop documentation regarding the implications of offering 3D printing services in libraries. Check the website at http://www.ala.org/offices/oitp for updates and additional reports.
- Parallel your library's and institution's existing policies governing reproduction equipment and user behaviors such as acceptable use and computer use.
- Complement your current public service policies regarding level of assistance, hours of service and expertise, and other related aspects.
- Don't reinvent the wheel! Check other libraries' policies for ideas to include or modify in your policy.
- Run your draft policy by your administration and legal team.
- Examine and alter your policy as needed. Some details may fit more appropriately in workflow procedures rather than in policies.

The next chapter outlines the workflow elements you are likely to use in your service, and provides examples from several libraries. It also explores some software tools to aid your workflow and tracking functions.

◎ References

Wapner, Charlie. 2014. "Progress in the Making: An Introduction to 3D Printing and Public Policy." Washington, DC: Office for Information Technology Policy. September 29. http://www.ala.org/offices/sites/ala.org.offices/files/content/3d_printing_tipsheet_version_9_Final.pdf.
———. 2015. "Progress in the Making: 3D Printing Policy Considerations through the Library Lens." Washington, DC: American Library Association. January 6. http://www.ala.org/offices/sites/ala.org.offices/files/content/3D_Library_Policy-ALA_OITP_Perspectives-2015Jan06.pdf.

Workflow

ESTABLISH A WORKFLOW FOR YOUR 3D PRINTING SERVICE as soon as possible, preferably before you roll out the service. The workflow will be tweaked early and often, but staff will feel more comfortable if they are armed with procedures and an understanding of responsibilities. Even a one-printer operation will provide more efficient service with a clear workflow, especially when requests begin to queue. Libraries with more than one printer will discover that the complexity of the service scales up rapidly. Software tools to aid both the workflow and the printing functions are being deployed with great frequency.

Staff-Mediated Printing: The Basic Workflow

The workflow for a staff-mediated service that has one printer should consist of the following basic elements:

- Initial questions and consultations
- STL files are submitted
- STL files are examined and prepared
- Patron approves
- Patron pays, if relevant

- STL files are processed into G-code
- Approximate finish time communicated
- Queue management
- Printing
- Notification, pickup, and closure

Initial Questions and Consultations

Initial questions and consultations can be managed outside of the printing workflow itself, but staff must be prepared for patrons who say they would like to 3D print something but who in fact are just at the beginning inquiry stage. All staff should know how to handle initial queries about the service, whether to give handouts, point to online guides, or refer to experienced staff. See suggestions for providing this level of staff training in chapter 9. The following examples illustrate how libraries gently hint that the entire process may flow more smoothly with a consultation that precedes the file submission stage.

Consultation Example: San Diego Public Library

The San Diego Public Library's 3D printing guide includes a recommendation for a consultation appointment:

To make an appointment:

1. Stop by our lab to get estimate on how long it'll take to print your object.
2. Contact us to make an appointment by email.

For large objects, we recommend that you stop by the lab first to get an estimate of how long it'll take, then make an appointment. We also recommend that you make your appointment at least one week in advance. (San Diego Public Library, 2015)

Consultation Example: Boise State University

At Boise State University, the Albertsons Library has created an online form that its institutional members may fill out in advance of a 3D printing consultation session. The form includes an option for requesters to specify their level of experience with 3D modeling. A similar form that queries about patrons' level of expertise with any aspect of your service can be helpful in guiding patrons to the appropriate staff or perhaps in estimating the time needed for a consultation. You may view the form on the library website (Boise State University, 2015).

File Submission and Acceptance

The STL files that contain your patrons' models can arrive via several channels. You may choose whether your patrons must submit STL files in person or electronically, or if you will permit several options for delivery.

If patrons must submit files in person, you should establish hours when files are accepted. You must also make available staff who can accept the files and at least prepare the job through the settings stage and to the point where the patron gives final approval.

You may choose to specify or be flexible whether patrons may bring their files on a thumb or USB drive, whether you will directly download STL files from an online website, such as Thingiverse.com, or whether you will allow patrons to use your processing computer to login to their e-mail or cloud storage to retrieve a file for downloading. Advantages to asking patrons for URLs to models found on online websites include not passing around large files, the opportunity for staff to read any instructions and hints for the model, and the chance to see a photo of the item and thus to make better decisions when preparing the file. The ease in accessing an STL file will vary depending on how the patron obtained or stored it. Permitting as much flexibility as your staff can handle in retrieving the file will result in the greatest convenience for your patrons.

If patrons submit files electronically, the preparation and processing stages can be handled when it is convenient for staff. Be sure to publicize estimates of when the patron can expect a reply with either further questions or notification that the job is processed and ready to print. Libraries accept files electronically through several routes. Some will only accept e-mail attachments, preferably not to a personal address, while others prefer not to rely on e-mail attachments. Some have created forms that permit the patron to browse and upload the file. Specify the file formats that you are willing to accept. Most libraries will only accept STL, OBJ, or .thing formats, but your options will vary depending on the printer software used. Be sure to specify the largest file size that you can or are willing to receive, whether you rely on e-mail attachments or upload options. If you have a generous file size allowance but encounter a patron with a model that exceeds your limit, you should suspect that the patron has a bad model that needs to be repaired; see chapter 4 before you explore options for managing an oversized file.

The McGoogan Library of Medicine at the University of Nebraska states a work-around on its 3D printing request form. "There is a 10MB file size limit as a file attachment. If your file is larger than 10MB, please bring it into the McGoogan Library on a flash drive for printing. You may also provide the link to the file in the Additional Information box below so we can download it" (McGoogan Library of Medicine, 2015).

Some libraries limit the number of STL files that may be submitted in a request. This can be considered a workflow management issue that is not necessarily linked to the policy issue governing the number of objects a library will print for one patron at a time (see chapter 7). Some objects are constructed from multiple pieces that must be printed separately for a variety of reasons (see chapter 4). One STL file upload per request may be a limitation of the uploading procedure rather than a limitation of the number of files that your staff can process at once. Balance the best interests of your patrons and your staff when placing limits on files per request. The University of Wisconsin, Stevens Point, University Library (2014) requests, "If you have several files to print, please submit each of these separately by filling out a separate 3D Printing Request Form for each print."

Some libraries request that patrons run their files through a model repair service, also known as model checking software (see chapter 4), before submission. This run-through is especially helpful for items that patrons have modeled themselves. Files that are found online may not require checking unless staff have a reason to be suspicious, such as for models that don't include a photo or acknowledgment that others have printed it successfully.

Libraries have created a variety of file submission procedures for their patrons to follow. Some examples, ranging from simple to more complicated, are included here.

Hinsdale (IL) Public Library

Step 3 of the Hinsdale [IL] Public Library's (2015) 3D print instructions at requests, "Attach the .STL file to an e-mail and send it to [e-mail]. Be sure to include your name and phone number. Projects may take a few days to print. Once yours is ready, we will contact you to pick it up in the Youth Services department."

Lake Forest Academy

The Lake Forest Academy is a grades 9–12 school. Its guide (Lake Forest Academy, 2014) states, "If you find a model online that you'd like to have printed, you may submit a link to that file with information. If you find a file online, it is helpful to us to send the link, not the file. The creator may have information about the file that we need to see. . . . You may submit an .stl file to be printed if you find or design a printable model. . . . Once you've read these guidelines, you are ready to submit your request [by e-mail]."

Lewes Public Library

The Lewes [DE] Public Library (2015) requests that a form be completed and that models be previewed before submission. Patrons may bring the printed form and a USB drive to the library, or e-mail the form along with a link to the file. One policy statement on the form notes that "jobs that require multiple printings may drop in queue position."

Sacramento (CA) Public Library

The Sacramento [CA] Public Library's (2013) procedure indicates the most basic type of file submission.

Submitting a design for printing:

1. Persons wanting to use the 3D printer shall bring their file (in .stl, .obj, or .thing file format) (no larger than 25MB) to the DesignSpot during open hours. Staff will add the model to the printing queue.
2. If there is high demand, the Library will schedule only one print per day per person or entity.
3. The files will be readied for printing in MakerWare or other authorized software. The Library will view all files in MakerWare or other authorized software before printing.
4. Wait/pickup time: Items may be picked up at the DesignSpot. It is sometimes difficult to estimate exact print times. Library staff will make an educated guess about the length of a job upon request.

Illinois Institute of Technology

The Illinois Institute of Technology, Paul V. Galvin Library (2015) posts the following instructions on its guide.

- Fill out the 3D Printing Request form
- Click on "I agree" to indicate your compliance with University and Galvin Library policies
- Upload your .STL file by clicking on the "Choose file" icon. Note: Only .STL files will be accepted

- You will receive a confirmation email when library staff receives your request form and file. If there is any additional information needed, it will be included in the email, along with an estimated time for when your print job will start. This email will also include an estimated cost for your print job.
- This is the only time in the process where you have a limited amount of time to CANCEL the print job before it is entered into the queue for printing.

*Be aware of time and printing requirements and *please* allow enough time for your print job. Most prints will take 2–12 hours, but some take longer and there may also be times when the printer is undergoing repair, is experiencing heavy use, or is being used for a special event.
*Course work will NOT be given special treatment or print times. Prints intended for course work will be entered as part of the everyday queue.

University of Wisconsin, Stevens Point

The University of Wisconsin, Stevens Point, University Library (2014) mixes policy and procedures on its guide. Procedures include:

Users must submit their files in .STL format. Users will need to fill out and submit the 3D Printing Request Form along with their .STL file. Library staff will review the file and send a confirmation email to the address provided that the submission has been received. The email will state whether the file has been approved, the estimated cost, expected completion data, and any important information for the user. Library staff may need additional information about the print job or may need to schedule a consultation with the user. Once the file has been printed, staff will send another email informing the user the cost of the print and the due date to pick up the model.

If you have several files to print, please submit each of these separately by filling out a separate 3D Printing Request Form for each print. . . .

BEST PRACTICES FOR FILE ACCEPTANCE

- Establish hours when you will accept files in person.
- Decide whether you will accept files in person, electronically, or both.
- Specify the file types you will accept, such as STL, OBJ, or .thing only.
- Balance staff and patron needs when specifying the number of files you will accept per submission.
- Consider requesting that patrons run their STL files though model repair software before submission.
- For models found in online libraries, check for instructions, hints, and photos to ensure successful file preparation.
- For in-person submissions, specify acceptable delivery media: portable drives, downloads, or both.
- For electronic submissions:
  - Specify a maximum file size you will accept.
  - Specify a maximum number of files you will accept per request.
- Permit sending URLs instead of files for models found in an online library.

All submissions are subject to approval based on scheduling and availability. There may be times that the printer is malfunctioning, being repaired, or is being used for an event or a course. During such times, the 3D printer may not be available for use and there will be a delay in approving submissions and printing objects. Any significant lapses in printing time will be noted on the 3D printing web page.

After the submission has been printed and the print has been picked up or the two week time limit to pick up the object is over, the submitted file will be deleted by library staff.

Several best practices emerge from the procedures listed above. They are summarized in the textbox.

File Preparation

After an STL file is opened in the 3D printing software, the model must be examined for printability (see chapter 4). Verification of the patron's intent for the size and scale of the finished object is a critical element in the workflow for two main reasons. Some objects are modeled in inches while most printers work in millimeters, and a simple conversion in the printer software usually solves that problem. Some patrons are unaware of the limitations of the printer's build size, and must be informed of the need to either rescale the model or re-model a large object into several smaller pieces.

Each object must be arranged on the plate in a position that ensures the most successful printing, as described in chapter 4. Efficiency in printing is a preferred goal but is not required. Experience and patience will enable staff and patrons to recognize the optimal object rotation for successful printing. Libraries may choose to oversee or "approve" this stage of file preparation to increase efficiency and also to reduce the likelihood of damage to the printers caused by ill-prepared jobs that cause filament to glob onto the extruder; see chapter 10 for further details of what can go wrong. Consider the odds of damage caused by inexperienced staff or patrons when establishing the level of file approval, recognizing that it may evolve as both staff and patrons gain experience with file preparation.

Other customizations that must be confirmed before the job is processed include the number of pieces needed of each object and the filament choice for each object. Printer-specific settings options such as rafts, supports, infill, the number of shells, and perhaps more must be negotiated or left up to staff discretion. These setting options affect the amount of filament used and the amount of time needed to print an object, and thus many libraries will require the determination of these variables before a job will be printed.

These variables require some level of communication with patrons. That communication can take place in person, via e-mail/chat/text, or through a submission form that includes as many of the options as possible. The file preparation process can be handled

---

BEST PRACTICES FOR FILE PREPARATION

- Verify scale/size of finished objects.
- Try several arrangements of objects on the build plate if needed to determine the most efficient and successful print job.
- Approve models and settings before agreeing to print.
- Communicate with patrons if the model or settings don't seem right.

in conjunction with the file delivery or separately, as your workspace and staffing permit. See examples of online forms for submitting requests in the section on "Software Tools to Track the Workflow" below.

Payment

Libraries that charge for 3D printing must determine the point at which payment is made. If your payment system requires an in-person transaction, it might make sense to choose an in-person file delivery workflow. If you accept online payment or if you charge to patrons' library or institutional accounts, you would not have to require face-to-face file delivery. Consider using or adapting any payment mechanism that you currently use to charge patrons for fines or other services.

File Processing

After the file preparation variables are set, the model is ready to be processed for printing. If the processing is not done immediately after the preparation variables are set, any changes to the patron's original submission (such as filament choice, weight, and time) must be communicated to the staff who will process the file. The processing stage can take a great deal of time, based on the size and complexity of the job. If you only use one computer for 3D printing, you may encounter situations when you cannot accept a new job until the previous job has finished processing. With in-person file delivery, a queue for drop-offs can become problematic for patrons who underestimate the time required to serve them and to serve your other patrons.

Your workflow must include a method of keeping the processed file, filament to be used, and a patron/job tracking element together throughout the process. At the University of Florida, staff fill out a slip of paper with job details (see figure 8.1), then tape the

---

Thank you for printing a 3D model!

Any questions? Go to guides.uflib.ufl.edu/3dprinter

Name: Card#:

File name: Color:

Est. grams: Est. time: Class? Y N

- -

taken by: date: time:

Figure 8.1. Example of a Job Processing Slip

SD card with the processed file to the slip, and preferably walk the slip over to the printer room. This slip is ultimately placed in a plastic bag along with the finished object when it is conveyed to the pickup location.

If you are tracking your 3D printing jobs for statistical purposes as well, the point after a file is processed is a good time to enter data into your tracking system because you now know approximately how much filament and time the print job will require. A simple spreadsheet may be all you need, and other software solutions are described below. Regardless of method, achieving staff compliance in entering data accurately and in a timely fashion requires training and vigilance.

Printing and Queue Management

Many libraries choose a "first come, first served" prioritization of print jobs. Some are careful to specify whether this prioritization is based on the times that requests are accepted, approved, or processed because complications and the time to resolve problems at any of these stages might reasonably push a job further down the queue than the patron originally expected.

The Illinois Institute of Technology (2015) states its prioritization in its guide on the library's website. "Files will be printed in the order that they are approved, not the order that they are submitted. An exception to this would be a small print job that would fit in the print time-frame available, printed ahead of another one in the queue before the printer stops for the day." The Hinsdale [IL] Public Library (2015) keeps the workflow process simple for its staff by declaring on its guide that "designs are one color only, and the color is determined by the filament loaded in the machine at the time of the request."

The printing operation itself is fairly quick and simple. Staff must check the printer for filament color and level, and unload the current filament and then load new filament if needed. Staff then send or place the processed G-code file in the printer, and press Start.

Queue management becomes a necessary function as soon as you have more than one job waiting to be printed. Staff will need to know which job to start next. Your system can be simple if you mark the date and time on slips or on the file transfer hardware. You may need to establish a queue (or production) manager who orders the queue and shifts the order as appropriate, based on any prioritizations such as user need or filament choice or length of job, throughout the day. If you have more than two or three staff involved in printing, you are likely to need someone to take the leadership role; otherwise, staff who are busy because the requests are pouring in will assume that a colleague is handling the other tasks. The queue manager may be the most appropriate person to check that data have been entered into your tracking system.

A 3D printer must be loaded with the requested filament before each job is started, and it can be challenging for staff to remember to check the color and amount of filament before pressing Start. Staff who are accustomed to changing out the thread on sewing machines will excel at this task. With experience, staff will develop the skill to estimate the amount of filament left on a spool and to judge whether the quantity is sufficient for completing the next job. For accuracy, obtain a scale to sit near the printer or filament stash. Staff can then weigh spools that are getting low on filament, subtract the weight of an empty spool, and mark the remaining filament weight on the spool. If the amount of filament needed for the next job is indicated on its slip, staff can compare that amount with the weight left on a spool and then know whether that spool has enough filament left for the job. Weigh some of your early pieces (with raft and supports still attached) so you can learn whether your printer software under- or over-estimates the amount of filament needed for typical jobs. Then you will know how to err when judging if a spool has enough for the next job.

When a job is completed, staff must remove it from the printer and, if appropriate, move it to the pickup location. If you choose to remove the raft or supports from printed objects, decide where and by whom that task will be performed. Many libraries will not remove supports for the patron, but it is easiest to remove the raft very soon after the job has completed. Rafts are easy to detach, but the supports may be delicate or not obvious and may be best left for the patrons unless they are children. If you have a way to recycle the used filament (see chapter 2), you may be more inclined to remove the raft yourself or to encourage your patrons to deposit the raft and support material in a recycling box.

Notification, Pickup, and Closure

You can notify patrons of the status of their print job at each stage: received, processed, in the queue, started printing, finished printing. Frequent notification adds to the workload of your staff. Even an automated method of informing your patrons will still require staff effort to update the system to trigger the note.

One option is to give patrons an estimated time of completion when the print job is approved. However, one of the most compelling best practices of 3D printing is to never promise or guarantee a completion time! Many bad things can happen (as outlined in chapter 10), ranging from a filament jam to a power outage, that require restarting a print job.

If you simply give an estimate, you will receive phone calls and e-mails as well as drop-in visits by patrons checking on the status of their jobs. Your staff must be able to respond to status queries by consulting your documentation or by looking at the queue.

BEST PRACTICES FOR QUEUE MANAGEMENT

- With a "first come, first served" prioritization, consider a "first processed" order rather than "first submitted."
- Consider adjusting the queue based on print time or filament choice.
- Remember to check the filament level before every job!
- Establish a queue manager if you have many staff, many jobs, or more than one printer.
- Establish a policy for raft and support removal.

You may notify patrons when their jobs are complete through several avenues. Simple options include phone calls, texts, or e-mails. You may prefer one method or you may ask the patron's preference when the job is submitted. If you use e-mail, consider whether to send the message from a staff member's individual e-mail or a collective library e-mail; your decision on how to communicate with the patron may depend on how you send notifications of other library services. An automated system can generate a notification when staff update the database.

When patrons pick up their jobs, the staff who hand off the item should ask for a name or ID to ensure that the item is going to the person who requested (and perhaps paid for) it. If your printer is in a visible area, problem patrons may attempt to claim that "the yellow dog that was printing yesterday" belongs to them. Occasionally patrons may ask, either at the time of submission or notification, if another person may pick up the item on their behalf. If you will permit proxies to pick up jobs, develop your procedures for indicating the permitted proxy in your workflow.

Patrons may have questions about cleaning up their finished pieces. Ensure that staff who hand out the pieces know where to direct the patrons for suggestions on tools or finishing products that are suitable for the type(s) of filament you offer.

For closure, update your records after a patron has picked up the printed item. This task belongs to the staff who hand out the item.

Patron-Initiated Printing: The Basic Workflow

With a patron-initiated service, patrons typically reserve time on the printer, process their own jobs, change out the filament as needed, and start the printer themselves. Below is a list of typical tasks for patrons to perform if managing their own jobs. Some of the tasks are identical to those described in staff-mediated service (above) but are performed by patrons and must be included in any training sessions provided for patrons. Staff may perform or provide oversight for some of these tasks. The general workflow for this service is similar to that of staff-mediated printing, with a few additions.

- Patrons receive (your library's) training.
- Staff add authorized/certified users to a list of approved patrons.
- List must be checked, by humans or by system, before patron can reserve a printer.
- List should be maintained/pruned as desired, perhaps annually.
- Patron makes or obtains model in STL.

- Patron makes all decisions regarding scale, quantity, infill, and other settings.
- Patron processes STL to G-code.
- Patron makes reservation.
- Patron may need to make sure the printing time does not exceed the reservation time limit.
- Patron might bring/buy own filament.
- Patron loads filament.
- Patron checks the build plate, applies tape or glue, calibrates/levels the build plate.
- Patron stays long enough to ensure success or for entire job.
- Patron adds note to a printer to indicate "in use" and maybe for how long.
- Patron must make sure every job (theirs or the previous) has completed before removing from build plate.
- Patron unloads filament.
- Patron can fill out logbook.
- Staff present or nearby to assist.

As workflow responsibilities shift from staff to patrons, some tasks become easier and some become harder to manage. For example, when patrons process their own jobs, staff will be less involved in the minutiae of file settings. However, these concerns are typically replaced by the task of managing reservations, including the limits on the amount of printer time available to a patron or the number of reservations permitted in a day or a week.

The concept of a reservation for 3D printer time is a tricky one. When something goes wrong with an earlier job (see chapter 10), or if a patron misjudges the time needed to complete a job, all of the subsequent reservations for that day will be thrown off course. The Tulsa City-County Library website (2014) states, "If you have already been to an orientation and signed the waiver for 3d printing, we ask that you call us ahead of time to set up an appointment to use the machines. This is so we can make sure a printer is available when you come in and not in use by staff or another patron."

You may need to establish and enforce some variations on the code of ethics for laundromat use. Possible examples are:

- Patrons may not pause or cancel other users' jobs.
- Patrons must contact staff if another user's job interferes with an appointment.
- Patrons may (or may not) remove a previous user's job if the printer indicates the job is 100 percent complete.

Below are examples that illustrate the workflow procedures established by selected libraries that offer patron-initiated printing.

Patron-Managed Workflow Examples at Public and School Libraries

Riverdale Collegiate Institute

The school library at the Riverdale Collegiate Institute (grades 9–12) in Toronto, Ontario, presents the procedures in a "rules" section on its handout (Dempster, 2015).

1. Every .stl design file must be sent to Netfabb 3D repair to "fix" the .stl file for printing.
2. You import your fixed .stl file into Makerware. You would generally do this on the library MacBook (the library has the SD card for the printer).

3. In MakerWare you make print choices (such as size and scale, which extruder nozzle to select, whether your design needs a raft or supports to print properly and more). The 3D Club members can help with those choices.
4. From Makerware, you create your final .thing file and load it onto the Makerbot SD card.
5. Finally, you need to arrange a time you can print it (including when staff is available for the time needed).
6. YOU MUST BE THERE TO SUPERVISE THE PRINTING PROCESS. This is a great time to do homework (as long as one eye is on the print!)

Westport (CT) Library

The public library in Westport, Connecticut, reminds its patrons on its guide (Westport Library, 2015) that the demand is high and time is limited.

At this time we have 3 3D printers and the public's demand for use is growing. This process will help to ensure that those wanting to use the printer at least one time, will have the opportunity.

While we make every effort to schedule as soon as possible, please ask for a time at least 48 hours in advance.

To schedule an appointment with a 3D printer coach for training, please check the online calendar for the current schedule of 3D printer coaches. The majority of these coaches are volunteers. Please select at least 2 days and times that you are able to come for training and send an email . . . for an appointment. (In most cases TWO 1 hour training sessions are required. Once the coach feels you are aware of how the 3D printer functions and how to solve problems when they occur, you can schedule a 2 hour session on the printer while a coach is there. Future times will be dependent upon the availability of printers.)

Patron-Managed Workflow Examples at Academic Libraries

University of Wisconsin, Stevens Point

The University of Wisconsin, Stevens Point, University Library (2014) provides both staff-mediated and patron-initiated 3D printing. Its guide explains, "If a user wishes to print their object themselves, they will need to schedule an appointment with library staff to receive training on the 3D printer. Users will be supervised by a library staff member during the printing process. The submission form will include this option and a library staff member will contact the user to schedule a training session."

University of Utah

The University of Utah (2015) provides extensive instructions for patrons on its guide.

When you are ready to begin printing, please follow these steps.

Purchase plastic filament from the Reserve Desk. Lengths of filament can be purchased at the Reserve Desk at $0.20/foot. You may also bring in your own filament so long as it is 3mm in diameter.

Create or find a 3D model in STL or OBJ format. Please see "Basic Information" and "Tips and Tricks" for basic information on model creation, sources for ready-made models, and 3D model file types.

Open the model in Cura, apply the appropriate Cura profile (config), and export to a G-code file. Cura profiles for a variety of filament types can be found and downloaded at https://www.lulzbot.com/support/taz-cura-profiles. As an alternative, you may also use Slic3r. Profiles for it can be found at https://www.lulzbot.com/support/taz-slic3r-profiles.

Open Pronterface, turn the printer on and connect to the 3D printer according to the instructions provided in the operating manual.

Heat the extruder, then load your filament according to the instructions provided in the operating manual. Please be aware that different materials require different temperature settings. That information can be found on the same page with the Slic3r profiles.

Load your G-code file and examine it in the "Preview" window. "Preview" allows you to see each printed layer of your model and check if there could be issues with the print. If there are, you will need to re-examine your model for non-manifold geometry, fix any issues and re-slic3 the model before proceeding. If all looks in order, start your print.

When your print is done, heat the extruder and unload your filament according to the instructions provided in the operating manual.

When the print bed is cool, remove your print. Please be aware that objects with large flat areas on the bed surface will be difficult to remove. Do not apply too much force when removing your model. Doing so could result in damage to your model or to the glass bed.

Turn the printer off.

University of Michigan

The University of Michigan provides "steps to walk-up 3D printing at U-Mich" at http://um3d.dc.umich.edu/3d-printing-walkup/, which includes instructions to review tutorials, take the knowledge test, reserve a printer, print, and clean up.

University of Memphis

The University of Memphis's guide "3D Printing @UofMLibraries: How to Print" at http://libguides.memphis.edu/c.php?g=94303&p=611087 advises patrons in how to participate in training, schedule lab time, and prepare files before (fill form, upload), during, and after printing.

BEST PRACTICES FOR PATRON-INITIATED WORKFLOWS

- Ensure patrons have received training, and only permit certified users to reserve the printer.
- Set up your reservation system to minimize problems.
- Assign staff to check on patrons and printers frequently.
- Assign staff to be readily available to offer assistance.
- Ensure that patrons handle only their own jobs.
- If using a tracking or notation system, ensure that patrons and staff make appropriate updates.

⚙ Added Complexities with Multiple Printers

When you add multiple printers to your fleet, the complexity of managing jobs increases exponentially. In the first place, patron demand is likely to increase for several reasons. Patrons may see that more printers are available, they will be encouraged if the queues are shorter or if the response to "Can I print right now?" is affirmative, and satisfied patrons will spread the word.

With staff-mediated printing, you are likely to require a queue manager. Accurate, complete, and timely record keeping becomes crucial when you run several printers. Maintaining an efficient production line requires regular attention. Your manager may choose to shift jobs among printers when any of a number of events occur that can potentially alter the queue, such as:

- When failed jobs have to be restarted (printing error) or scrapped (bad model).
- When a problem printer requires special attention to get it working again.
- When unusually brief or long jobs enter the queue.
- When one spool of filament is needed for more than one job at a time.
- When you offer several filament choices, you will gain efficiency and reduce some potential printing problems by grouping jobs that use the same filament.

The greater your staff size, the more likely you will need a queue manager. You will also need to ensure that printing staff report any problems they encounter, whether to peers or to the manager. Your communication system can consist of sticky notes, individual or group e-mails, your tracking database (see below), or several of these methods—whatever results in the greatest compliance by your staff.

With patron-initiated printing, the management of your reservation system (whether low tech or high tech) will involve more effort. Communication with patrons about printer availability will become more complicated, with the added tasks of confining patrons to their reserved printers only. Becoming aware of printer problems and solving them will become more challenging, although your skill set will increase and the time to solve a particular problem will decrease with repetition. Record all printing problems so patterns of failure can be observed and treated. The queue manager should report problems as soon as possible to the troubleshooter/maintainer if that is a different person.

The Fayetteville [NY] Free Library (2014) notes its methods for efficient distribution of jobs among its printers on its FAQ. "There are occasionally times when our makerspace is very busy and all 5 3D printers and laser cutter are in use. To resolve this issue, we have put several new procedures in place. We limit each patron to using one 3D printer at a time. We also allow 2 of 3D printers to be reserved each day starting 3 hours prior to close. That way, if a patron comes in wanting to use a printer and they are all taken, they can at least make a reservation for the next day or a later date and know they will have guaranteed access."

The University of Michigan's form at http://um3d.dc.umich.edu/3d-printing-submit -a-model/ includes an option to choose the type of plastic desired. The filament selection governs which printer will be used for each job.

Software Tools to Track the Workflow

You will need to manage and track your jobs somehow, whether by paper, simple spreadsheets, or more complex software solutions. Plan to include the elements needed both to manage the workflow and to provide you with assessment data. Fields for spreadsheets, databases, or paper-based tracking may include:

- Date/time job was received
- Filename or tracking number
- Patron contact information
- File location (such as card or USB, if helpful in addition to filename)
- Date/time job was approved
- Grams and minutes estimated for the job
- Filament (if more than one option)
- Printer (if more than one)
- Name/initials of staff who accepted/approved the job
- Date printed
- Date patron was contacted
- Date of pickup
- Notes to record do-overs and unusual aspects

You may find it helpful to devise a tracking slip that stays with each job throughout the process (see figure 8.1 earlier in this chapter). Explore whether such a slip can be generated from your spreadsheet or database. For low-tech shops, try to reduce the redundancy by not overrepeating data in your formal tracking and in your job tracking. When staff have to record the same data in too many places, errors or omissions are more likely to occur.

With an online form for patrons to submit jobs, the data can be captured into a database or spreadsheet for use throughout the workflow and for long-term recording. If you restrict the service to your user community, explore whether your online submission form can simply request the institutional user ID or a library card number and then populate the form with contact information derived from your patron database.

Examples of online submission forms can be found at:

- Atlantic City Free Public Library http://make.acfpl.org/3d-printing-request-form/
- Boise State University http://guides.boisestate.edu/c.php?g=74312&p=474642

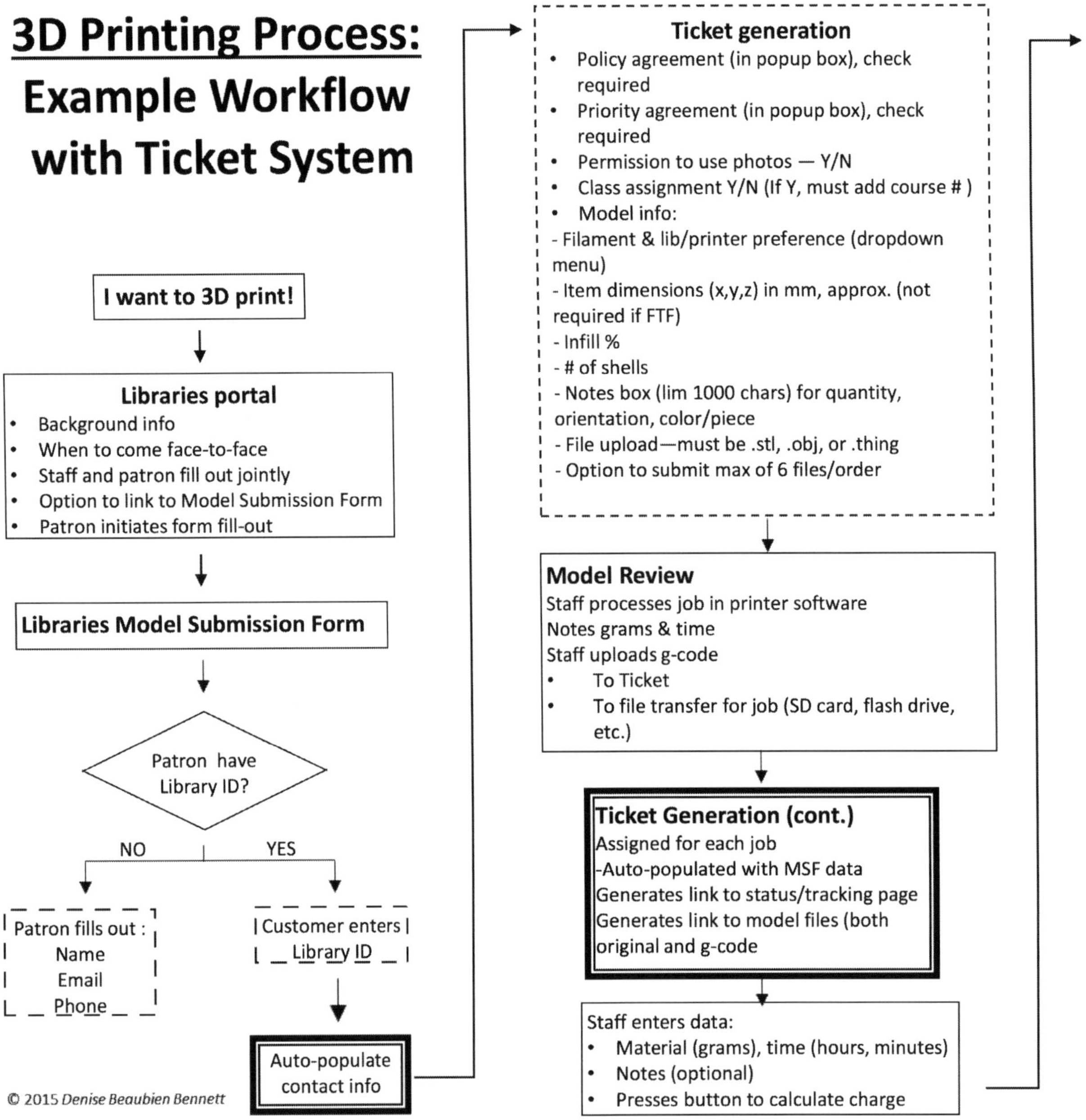

3D Printing Process: Example Workflow with Ticket System

I want to 3D print!

↓

Libraries portal
- Background info
- When to come face-to-face
- Staff and patron fill out jointly
- Option to link to Model Submission Form
- Patron initiates form fill-out

↓

Libraries Model Submission Form

↓

Patron have Library ID?

NO ————— YES

NO →
Patron fills out :
Name
Email
Phone

YES →
Customer enters
Library ID

↓

Auto-populate contact info

© 2015 *Denise Beaubien Bennett*

Ticket generation
- Policy agreement (in popup box), check required
- Priority agreement (in popup box), check required
- Permission to use photos — Y/N
- Class assignment Y/N (If Y, must add course #)
- Model info:
- Filament & lib/printer preference (dropdown menu)
- Item dimensions (x,y,z) in mm, approx. (not required if FTF)
- Infill %
- # of shells
- Notes box (lim 1000 chars) for quantity, orientation, color/piece
- File upload—must be .stl, .obj, or .thing
- Option to submit max of 6 files/order

↓

Model Review
Staff processes job in printer software
Notes grams & time
Staff uploads g-code
- To Ticket
- To file transfer for job (SD card, flash drive, etc.)

↓

Ticket Generation (cont.)
Assigned for each job
-Auto-populated with MSF data
Generates link to status/tracking page
Generates link to model files (both original and g-code

↓

Staff enters data:
- Material (grams), time (hours, minutes)
- Notes (optional)
- Presses button to calculate charge

Figure 8.2. 3D Printing Process Example Workflow. *Copyright by Denise Beaubien Bennett*

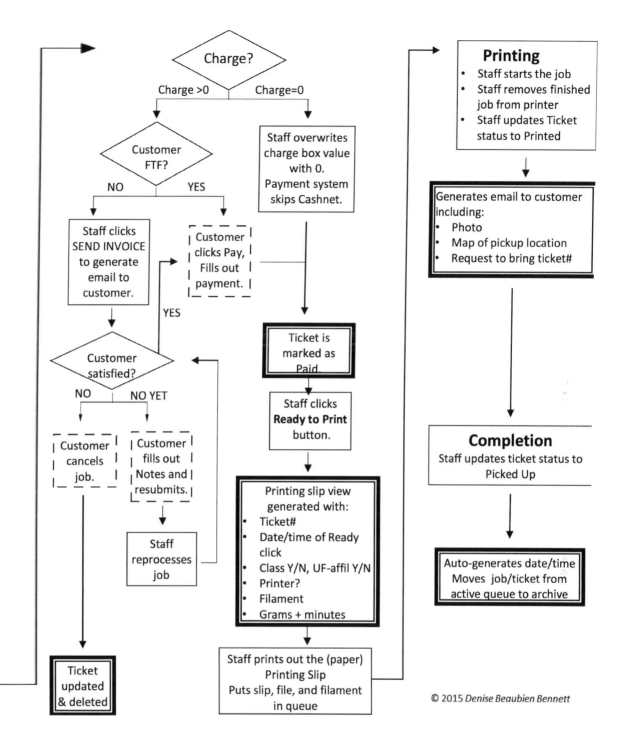

Charge?

Charge >0   Charge=0

Customer FTF?

NO   YES

Staff overwrites charge box value with 0. Payment system skips Cashnet.

Staff clicks SEND INVOICE to generate email to customer.

Customer clicks Pay, Fills out payment.

YES

Customer satisfied?

NO   NO YET

Ticket is marked as Paid.

Customer cancels job.

Customer fills out Notes and resubmits.

Staff clicks **Ready to Print** button.

Staff reprocesses job

Ticket updated & deleted

Printing slip view generated with:
- Ticket#
- Date/time of Ready click
- Class Y/N, UF-affil Y/N
- Printer?
- Filament
- Grams + minutes

Staff prints out the (paper) Printing Slip
Puts slip, file, and filament in queue

Printing
- Staff starts the job
- Staff removes finished job from printer
- Staff updates Ticket status to Printed

Generates email to customer including:
- Photo
- Map of pickup location
- Request to bring ticket#

Completion
Staff updates ticket status to Picked Up

Auto-generates date/time Moves job/ticket from active queue to archive

© 2015 *Denise Beaubien Bennett*

- Illinois Institute of Technology http://form.jotform.us/form/42194005392147
- University of Arizona http://new.library.arizona.edu/visit/print/3D/request
- University of Maryland http://makerspace.lib.umd.edu/forms/3dprint.php

Case Studies: Examples of File Submission Workflows

The case studies below illustrate the precise steps used by three libraries to track their 3D printing submissions. The level of detail may prove useful to other libraries that wish to develop their own solution to managing their print jobs.

Boise State University provides a description of its file submission practices in a case study presented at the end of this chapter. The case study illustrates a typical evolution in recording systems as the number of print requests grows.

At North Carolina State University, Adam Rogers developed a database system for managing the printing queue at the staff-managed Hunt location (Rogers, 2015). By relying primarily on the status of jobs, the system has remained flexible.

I built a 3D printing queue system for the Hunt Library Makerspace at NC State University. While the system is essentially a web database, and could have been done any number of ways, such as with MySQL and PHP, I used a Software as a Service (SaaS) solution called Knack (http://www.knackhq.com) which allows for simple creation of database-driven web apps in a style similar to Google Docs or Zoho. In this way, I leveraged the understanding of databases I had from library school, but did not have to get server access or do any coding (not my expertise). We pay a monthly fee for Knack, but it is reasonable ($20–30/month with an education discount).

Our system has two types of data: users and files, with everything else related in the metadata or attributes of files, (e.g. printer type, filament color, amount of plastic used, cost). The key to our printing queues is the "Status" attribute which defines where the file is in the printing process. Our system is mostly a series of different views on a list of files, filtered by Status so we can see easily everything that, for instance, is waiting to print or is currently printing. Another feature I built into our system is when a file reaches "Ready for Pickup" status, the user attached to that file gets an automated email telling them so; this was made using Knack's "Rules" feature.

This in-house system has been working well for over two years now, and has been flexible enough to accommodate changes in service workflow, new printers, price changes, and more. We have run into space issues at times because we do temporarily store users' STL files, which can take up space; other than that we have had very few issues.

The University of Florida partnered with its campus web designers to develop a queue management system for a patron-charged, staff-managed 3D printing service. Librarians developed a flowchart (see figure 8.2) to depict the steps and roles involved in the process in an effort to improve an inefficient workflow that required staff to record data in an online payment system, a spreadsheet, and on paper slips that move with each job. The new system eliminated redundancy in recording and provides more accurate statistical tracking of the service.

Software Tools to Assist the Printing Function

A model passes through several pieces of software before it emerges as a 3D printed object. Figure 8.3 illustrates this trail.

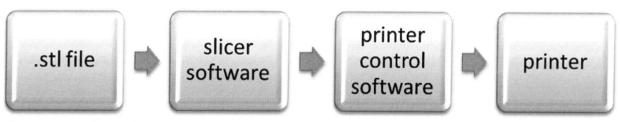

Figure 8.3. Software Flow

3D printing software performs two major tasks: one is to process the 3D model file for printing, and the second is to control and calibrate the printer. Processing software takes the desired parameters (e.g., infill percent, resolution, number of shells) and creates a set of instructions (sometimes called "G-code") for the printer to follow to extrude the 3D model. The slicer also provides time and filament amount estimates. This is commonly known as the slicing program and can be stand-alone or bundled with the printer control software.

Each 3D printer manufacturer will either specify the proprietary slicing software needed or will recommend one or more nonproprietary options. Many of the slicing programs, whether open source or proprietary, are free to download by staff or by patrons. Since patrons may use free software, libraries can permit or require that patrons process their own files to reduce some of the staff workload.

Printer control software allows the user to manage the 3D printer, including leveling, loading/unloading the filament, and calibrating the printer's movement. This software is often bundled with slicing software for a full printer management solution.

If your staff are managing multiple printers from different manufacturers, you should definitely consider both slicing and printer control software that interfaces with all of your models. It will save the sanity of both staff and patrons.

Table 8.1 outlines a variety of software programs available in the summer of 2015. It includes both slicer software to process models and printer control software that can manage one or more printers and in some cases also slice (or reslice) models. Both open source and commercial tools are available, and some also provide functionality for job tracking. Open source options can be customized locally, and some commercial vendors may be amenable to further development to suit your needs for a fee. Perhaps some libraries will begin sharing or marketing their homegrown solutions. Additional software programs are listed and updated at 3dprintingforbeginners.com (2015).

🌀 Key Points

Set up a workflow and test it before trialing your service. Your workflow will evolve as your staff gain experience. Best practices for ensuring a smooth 3D printing workflow include:

- Choose a different workflow for staff-managed and patron-initiated services.
- Establish distinct workflows for the patron interaction and production management functions.
- Brace for significant increases in workflow complexity if you add another printer to your fleet.

Table 8.1. Slicer and 3D Printer Control Software

NAME	COST AND LICENSE	PRINTER COMPATIBILITY	NOTES
Slicer Software			
Slic3r http://slic3r.org	Free, open source	Most printers (see list at http://slic3r.org/)	Can slice for multiple extruders
Skeinforge http://fabmetheus.crsndoo.com/	Free, open source	Many	Python scripts for slicing
Printer Control Software			
AstroPrint https://astroprint.com	Free plan, Pro plan available, AstroBox for wireless printing $149	Printers that accept G-code	Can slice and store 3D files, control and monitor printers in the cloud.
BotQueue http://botqueue.com	Free, open source	RepRap printers	Will reslice a model if needed for a different type of printer.
Cura http://software.ultimaker.com/	Free, open source	Developed by Ultimaker, RepRap printers	Slices and controls printer, LulzBot has custom edition of Cura
MakerBot Desktop https://www.makerbot.com/desktop	Free, proprietary	All MakerBot Replicators	Slices and controls printer, generates .x3g, .makerbot
OctoPrint http://octoprint.org/download/	Free, open source	List at https://github.com/foosel/OctoPrint/wiki/Supported-Printers	Web interface, connects wirelessly to printer using Raspberry Pi
PrintToPeer http://www.printtopeer.com/	In Beta testing (Summer 2015)	RepRap printers, MakerBot	Slices and controls printer
Pronterface http://www.pronterface.com/	Free, open source	RepRap printers	Part of Printrun suite
Repetier-Host http://www.repetier.com	Free	RepRap printers	Uses either Skeinforge or Slic3r
ReplicatorG http://replicat.org/	Free, open source	RepRap printers, older MakerBot Replicators	
Simplify3D https://simplify3d.com	$149 per license	List at https://www.simplify3d.com/software/supported-printers/	Slices and controls multiple printers, can repair mesh errors
Skyforge http://skyforge.co	Annual contract with educational discount for libraries	Many	Slices and controls multiple printers, can also manage your payment system

- Assign a queue manager if you have several staff or more than one printer.
- Simple paper and spreadsheet tools may be all you need to manage your workflow.
- Software tools can help manage your workflow. You can develop effective tools in-house, use open source (free) solutions, or license commercial programs as your skills, complexity, and budget permit.

In the next chapter, you will find recommendations for preparing your staff to accept the new service as well as tips for developing an effective staff training program.

Appendix
Case Study: Boise State University: Handling File Submissions

A query posed to the Librarymakerspace-L discussion list titled "How does your library handle file submissions" generated a response and follow-up e-mail by Deana Brown and Amber Sherman of Boise State University (Brown and Sherman, 2015). Below is a slightly modified version of their responses.

Here at Boise State University, the 3D Printing Team (4 of us) has been using a similar model for our 3D printer service (see http://guides.boisestate.edu/3dprinter) for a little over a semester. We started off with a simple Google Form (see https://docs.google.com/a/boisestate.edu/forms/d/1P5gpV7KnskGWM-9GeO3JLtEWgkaxixC_9_A8vprmmgc/viewform), but were struggling with tracking issues and figuring out how to manage files. It captured a good chunk of information, but file submission was still done through e-mail and with 4 of us running the service, all of our inboxes began to overflow!

We investigated alternatives, but most had an associated cost, which we were trying to avoid. Amber Sherman, our Scholarly Communications librarian, suggested we take a look at ScholarWorks from Digital Commons.

Though it isn't designed for 3D printer file submission / queue management, we have been using the form at http://guides.boisestate.edu/c.php?g=74312&p=474642 for this purpose since March. One of our Scholarly Communications librarians has been working closely with bepress to tweak their system to better fit our specific needs. Though you can't send print jobs to the printer through it, it is a way to track use and partially automate communication with users. We still have to use a Google Doc to determine print order, as we haven't figured out a good way to do this in ScholarWorks, yet. It's definitely a work in progress, but much better than our original system, and something we were already paying for.

Then we moved onto a combination of ScholarWorks (submission/automation of communication to users), and a Google sheet (queue management). My hope is to do away with the Google sheet eventually, and just communicate print queue information at the beginning of the week via e-mail.

Fields in ScholarWorks—File Submission

- Title/description of object
- Name/e-mail
- Department/major

- Is this for a class?
- Due date
- Color
- Notes
- Upload file
- Print status (staff use only)

Fields on Google Sheet—Printer Queue

- Name
- E-mail
- Status
- File name
- Quantity needed
- Quantity printed
- Color
- For class / due date
- Estimated print time
- Notes
- Hours
- Filament used
- Other sheets—Files need to be reworked, done, waiting for files

We have actually found those things [setting up prints, preheating, changing filament] to take less time than helping users learn about 3D modeling/printing! We spend a lot of time communicating via e-mail about ways users can make their files printable, and troubleshooting when the printer goes down.

Ideally, [our staff workflow is] an e-mail [that] has been sent out the previous night with a printing plan that includes who is printing what, and other details about the prints. When this doesn't happen, the first person in has a look at the queue, and prints the next thing in line. We're pretty good about keeping each other in the loop, and have a good idea if there are priority prints, or ones we're waiting for redesigns of. Prints for class assignments get priority, but we do not advertise this fact. Until everyone was trained to the same level, we also played to our strengths by dividing up slicing, printing, and queue management responsibilities.

◎ References

3dprintingforbeginners.com. 2015. "3D Modeling Tools." Software and Tools for 3D Printing. Accessed August 10. http://3dprintingforbeginners.com/software-tools/.

Boise State University. 2015. "Albertsons Library 3D Printer Consultation Request." Boise State University. Accessed August 10. https://docs.google.com/a/boisestate.edu/forms/d/1P5gp V7KnskGWM-9GeO3JLtEWgkaxixC_9_A8vprmmgc/viewform.

Brown, Deanna, and Amber Sherman. 2015. E-mail message to the authors, May 27.

Dempster, Lisa J. 2015. "Riverdale CI Library 3D Printing." RCI Library, Riverdale CI. Accessed August 10. https://goo.gl/w83h8W.

Fayetteville [NY] Free Library 2014. "Makerspace FAQ: 21 Questions about FFL Makerspaces." Fayetteville Free Library. http://www.fflib.org/make/makerspace-faqs.

Hinsdale [IL] Public Library. 2015. "3D Print." Hinsdale Public Library. Accessed August 10. http://hinsdalelibrary.info/how-do-i/3d-print.

Illinois Institute of Technology, Paul V. Galvin Library. 2015. "3D Printing." *Research Guides at Illinois Institute of Technology*. Paul V. Galvin Library. June 8. http://guides.library.iit.edu/3D Printer.

Lake Forest Academy. 2014. "3D Printing at LFA." *LibGuides at Lake Forest Academy*. Lake Forest Academy. May 20. http://lfanet.libguides.com/content.php?pid=560139&sid=4617185.

Lewes [DE] Public Library. 2015. "Lewes Public Library 3D Printing Request Form." Lewes Public Library. Accessed August 10. http://www.leweslibrary.org/files/3D%20Printer%20 form.pdf.

McGoogan Library of Medicine. 2015. "3D Printing Request Form." McGoogan Library of Medicine. University of Nebraska. Accessed August 10. https://unmc.wufoo.com/forms/3d -printing-request-form/.

Rogers, Adam. 2015. E-mail message to the authors. September 14.

Sacramento [CA] Public Library. 2013. "3D Printer Policy and Procedure." Sacramento Public Library. August 22. http://www.saclibrary.org/About-Us/Policies/3D-Printer-Policy-and-Procedure/.

San Diego Public Library. 2015. "Maker Lab." San Diego Public Library. August 4. http://sandiego .communityguides.com/MakerLab/.

Tulsa City-County Library. 2014. "3D Printing at the Library." June 25. Tulsa City-County Library (blog entry by Peter). http://www.tulsalibrary.org/blog/3d-printing-library.

University of Utah. 2015. "3D Printing Services: Do-It-Yourself Printing." July 15. http://campus guides.lib.utah.edu/c.php?g=160793&p=1455514.

University of Wisconsin, Stevens Point, University Library. 2014. "Policy and Procedures for 3D Printing at UWSP Library." University of Wisconsin, Stevens Point. July 28. http:// libraryguides.uwsp.edu/content.php?pid=603794&sid=5066816.

Westport Library. 2015. "MakerSpace: 3D Printers." Westport Library. Accessed August 10. http://westportlibrary.org/services/maker-space/3d-printers.

CHAPTER 9

Preparation and Staff Training

IN THIS CHAPTER

▷ Addressing staff concerns

▷ Developing staff buy-in

▷ Testing behind the scenes

▷ Training staff

▷ Developing training documentation

THIS CHAPTER WILL GUIDE YOU IN PREPARING your staff to meet the challenge of providing 3D printing to patrons. To ensure success, libraries must engage their staff early in the service development, provide adequate staff training, and choose the depth of expertise they will offer in each related service. As you develop staff training materials, consider which of them can also be used for patron training and possibly as certification materials as well, and devise the documents to suit as many audiences as possible.

Addressing Staff Concerns

You may already have an established approach to addressing the concerns that staff tend to express when any new service is added, such as concerns that they will have to do more work, that they don't know how to do these new tasks, or that they don't want to learn about this service or its associated tasks.

As discussed in chapter 5, there is a staffing cost to adding a new service. 3D printing will likely take more staff time than anticipated, so you must either construct a plan to reduce time spent in other tasks, hire more staff, add volunteers, or otherwise assist your staff in reprioritizing their assignments. Most of the staff time will be taken by those who interact with the public.

Staff may also be concerned about learning new tasks. These concerns can be addressed through the promise and delivery of a sound staff training program. The needed elements are discussed in the "Training Staff" section below.

A 3D printing service is likely to generate additional concerns. Staff may wonder why the library is adding this service, and may be concerned that the associated tasks differ significantly from their current job responsibilities. Chapter 1 offers some examples of the benefits of a 3D printing service and of its compatibility with the missions of many libraries. You should have developed a response to the "why?" question, perhaps adapting those of other libraries, in the very earliest stages of considering the service and identifying the funding sources. If not, the response should become apparent while you are drafting the policies that align with your library's or institution's mission; see chapter 7 for policy examples. Make sure you are communicating your library's rationale early and often to your staff so they can understand and accept the idea before the printer arrives.

Many aspects of a 3D printing service are unlike any that have been assigned to staff in the past. In the best case, you will have enough staff who are interested or who quickly realize that the needed skill set is possible to learn. If needed, you can reassure staff that they are not on the bleeding edge of providing a library 3D printing service. Many other public and academic libraries have succeeded and are happy to share their expertise with new installations.

Some libraries expect that each of their staff, especially those who work with the public, should be trained and capable of supporting the 3D printing operation. For example, the Fayetteville [NY] Free Library's Makerspace FAQ (2014) states, "Every member of our professional staff (8 librarians) has one 2–3 hour shift/week on the Fab Lab Help Desk. After 90 days of employment, our support staff members are also trained on the Fab Lab Help Desk and each have one 3–4 hour shift each month. This time is then supplemented with volunteer staffing of the Lab." You will not be in an unusual position if you plan to incorporate some aspect of the service into each staff assignment. Take care to offer service only during hours when adequately trained staff are available. Consider altering your staffing pattern if patron demand is poorly accommodated by your initial schedule.

Developing Staff Buy-In

It is never too early to begin preparing staff once a library has decided to explore the possibility of starting a 3D service. The more staff participation, the better the service will be received and rolled out to your patrons. This is especially true for participation from circulation, reference, and IT units who can provide input and testing. 3D printing is still an unknown technology to many, and substantial training and exposure may be necessary to demystify it for staff.

Be prepared to receive basic 3D questions from the public as soon as word begins to spread that the library may offer 3D printing. Staff can get a jump start on training before the printer arrives by watching manufacturer videos, downloading freely available printer software, and visiting nearby 3D print labs.

As with all new services, staff will respond with a range of enthusiasm and interest. It is difficult to predict who will express the most interest in playing with the new equipment, but staff who enjoy tinkering or creating things will find similarities when working with the 3D printers. For example, an FFF printer bears a strong resemblance to a sewing machine, since both devices feed a material of specific thickness and color from a spool through the machine.

There are a variety of opportunities for contributing to a 3D service, and staff may choose to specialize if given the option. The ability to visualize in three dimensions is not a universal skill across all library staff, and may not be easily acquired by those who do not possess the skill naturally. Figure 9.1 shows an example of the skill needed to process 3D print jobs successfully. In this figure, do you think the objects on the left are overlapping? The right side shows the under view, and reveals that the objects do overlap and would print as one piece if not moved in the printer software. Knowing to rotate the view to visualize overlapping components is a skill that may not be easily learned by all staff.

Staff who have difficulty with the visualization aspects of moving, turning, and scaling objects in the printer software will struggle with the patron interaction and model processing tasks. Some staff will have a greater aptitude and comfort level than others in disassembling the printer when needed. Most if not all staff should be capable of the most basic tasks of removing completed jobs from the printer, changing filament, and starting the next job. To successfully implement a new library service, it is important to customize your workflow to meet the strengths of your staff and make sure that all have the opportunity for sufficient training. Some staff will learn quickly while others will require repeated shadowing to develop a level of comfort with the software and equipment. The bottom line: with a staff of any size, do not expect that everyone will perform or learn all of the skills involved in a 3D printing service, and assign tasks accordingly. Figure 9.2, "Pyramid of Expertise for 3D Printing Assignments," illustrates that you can expect all of your staff to understand the basics, but with each layer of expertise needed, fewer of your staff are likely to acquire the skill set needed to become expert at the task.

Your patrons will notice if your staff are undertrained to support the service. On-the-job training is acceptable to all but the most impatient patrons, especially when knowledgeable staff are teaching the patron as well as the colleague. Experienced patrons who

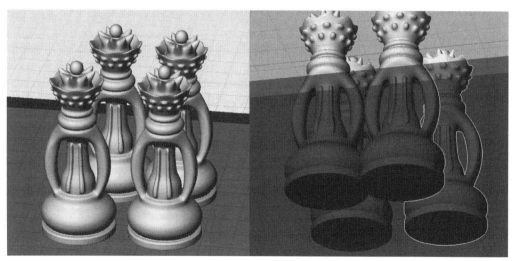

Figure 9.1. Visualizing in 3D

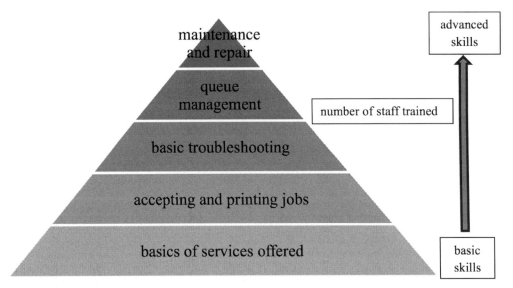

Figure 9.2. Pyramid of Expertise for 3D Printing Assignments

are still excited about 3D printing are likely to willingly assist with point-of-use training for staff or for other patrons. You must, however, establish options for untrained staff to request help from a more experienced colleague, or to receive sufficient training to provide quality service unaided as your service increases. Do not assume that initial training will be sufficient for the long run. Staff at each level of expertise will need to keep growing their skill sets as the technology matures and patrons everywhere develop increased expectations of 3D printers and of their owners and managers.

Testing behind the Scenes

When the printer arrives, you should set it up in a staff area and run a few jobs out of the sight of admiring patrons to reduce staff anxiety or embarrassment. If you can identify a patron or two who are knowledgeable about 3D printing and ready to print specific items, consider enlisting their help as your beta testers. A necessary qualification in a beta tester as well as in the lead staffer is patience. Similar to making pancakes, your first attempts are likely to be flawed while you learn how to level the build plate, determine suitable heating temperatures, remove finished pieces from the build plate, and gain confidence in removing rafts and supports from the objects. Offer free or exclusive first jobs to your beta testers in exchange for their assistance and patience.

The beta testing stage gives you an opportunity to test-drive your workflow (see chapter 8) in addition to learning how to operate the printer. This is an ideal time to engage the staff. They can try out and refine the workflow, developing familiarity with the procedures at this point.

Don't wait too long before rolling out the printer for general service, since your patrons (and perhaps your administrators) are likely to be clamoring for its debut. You may not have the luxury of waiting until many staff feel truly comfortable with the service and the workflow, but make sure you appear competent on your launch day. Further refinement of your workflow will continue as you encounter new and unexpected service challenges.

⊚ Training Staff

A staff training program should include the following phases of service:

- Services offered
- Print job acceptance
- Print job processing
- Printer operation
- Queue management
- When to call in the experts or suggest a consultation/appointment
- Troubleshooting
- Maintenance and repair
- Notification, pickup, and closure

Not every staff member will need to be trained in each phase. Make sure that each staff member receives hands-on training in the phases relevant to their assignment. Mercifully, you don't need to have all of the procedural components fully developed before you begin staff training. But keep an eye out for which elements should be included for the long term and for continuous training of current and future staff.

The Fayetteville (NY) Free Library states in its Makerspace FAQ (2014) the expectation that all staff participate in activities, and that they receive initial and ongoing training. This example of expectation plus training can serve as a model for other libraries.

> Staff training is an ongoing process as our makerspace activities grow, change and evolve. Our goal is to have all members of staff operate with a basic knowledge of the machines and equipment available in our spaces. We train all professional members of staff to be able to provide one-on-one or small group training sessions on the equipment such as 3D printing and laser cutting. We have monthly staff "lunch and learn" opportunities for ongoing training. Our staff team collectively decides what training sessions we would like to see offered or repeated during our once monthly "Maker Forums."

If you run a patron-initiated printing service or option, you can develop training items that serve both your staff and your patrons. For example, the Gloucester County [NJ] Library System (2015) presents a series of quickstart guides and handouts on its website for easy viewing. Likewise, the Fayetteville Free Library provides videos (2013 and 2015) to assist both staff and patrons. The University of Alabama (Scalfani, 2015) posts a guide detailing operating procedures for patrons and staff. Posting selected training items on your public website can empower and inform your patrons as well as your staff, enabling your 3D printing transactions to proceed more efficiently and with less stress on everyone's part.

Services Offered

All frontline staff must know the most basic elements of your service as described in chapter 6, and should be able to answer the following questions at least minimally:

- Do you offer just 3D printing? Do you have any 3D modeling software or services? Do you have any 3D scanning equipment for in-library use or for checkout?
- May I use the printers directly, or do your staff handle the printers?

- Does it cost anything?
- During what hours may I submit a file, use the printer, or consult a specialist?
- How do I make an appointment or reservation, if used?
- Is there a wait to use the printer? How long does it take for my job to print?
- How do I get started? Do you offer training or workshops?
- Where do I find a model?

These questions should be addressed in online guides and printed handouts for patrons. Staff who are less familiar with the service can then turn to those aids when responding to queries.

Print Job Acceptance

All frontline staff should know the fundamentals of your workflow and where to direct patrons with queries. Staff should also know which colleagues are responsible for which components to make efficient referrals. Staff who perform the basic workflow functions should know how to do the following:

Accepting Jobs

With patron-managed printing, most staff may not need to be trained in much beyond directing patrons to the printers or to the reservation system. With staff-mediated printing, staff training should include:

- Acceptable model formats (such as STL or OBJ)
- How to turn, rotate, and resize models
- Negotiating variables such as resolution, temperature, infill, and shells
- Recognizing bad models and suggesting options for repair

Your printer manual or the software user guide should include the instructions needed to perform most of these functions. However, most manuals are too lengthy to be helpful at the point of need. Copy and paste the relevant steps into a quick guide for your staff, adding in your specific workflow steps where appropriate. Use pictures instead of words wherever possible. Figure 9.3 illustrates how the University of Florida Libraries captured an image of the MakerBot software and superimposed the order of steps to be followed. It includes images from the printer's instructions that have been augmented with circles, arrows, and numbered steps to quickly illustrate what staff need to do when beginning to accept a print job. Inexperienced staff can easily consult the image, even during a patron consultation, and proceed smoothly.

You may not be able to establish "best practices" for your printer and users at the beginning stage of staff training. Your list of best practices will grow as you gain experience with accepting jobs and as you understand how your printer really works in your setting. Best practices may be local and are not always transferable; for example, when to recommend or discourage a particular resolution setting will depend highly on your printer's performance. Be sure to establish and maintain a list of best practices in your staff training manual. Some best practices for accepting and processing jobs are listed in the textbox.

The University of Alabama Libraries embed best practices in the Standard Operating Procedures document (Scalfani, 2015) that serves as a quick guide for users of its

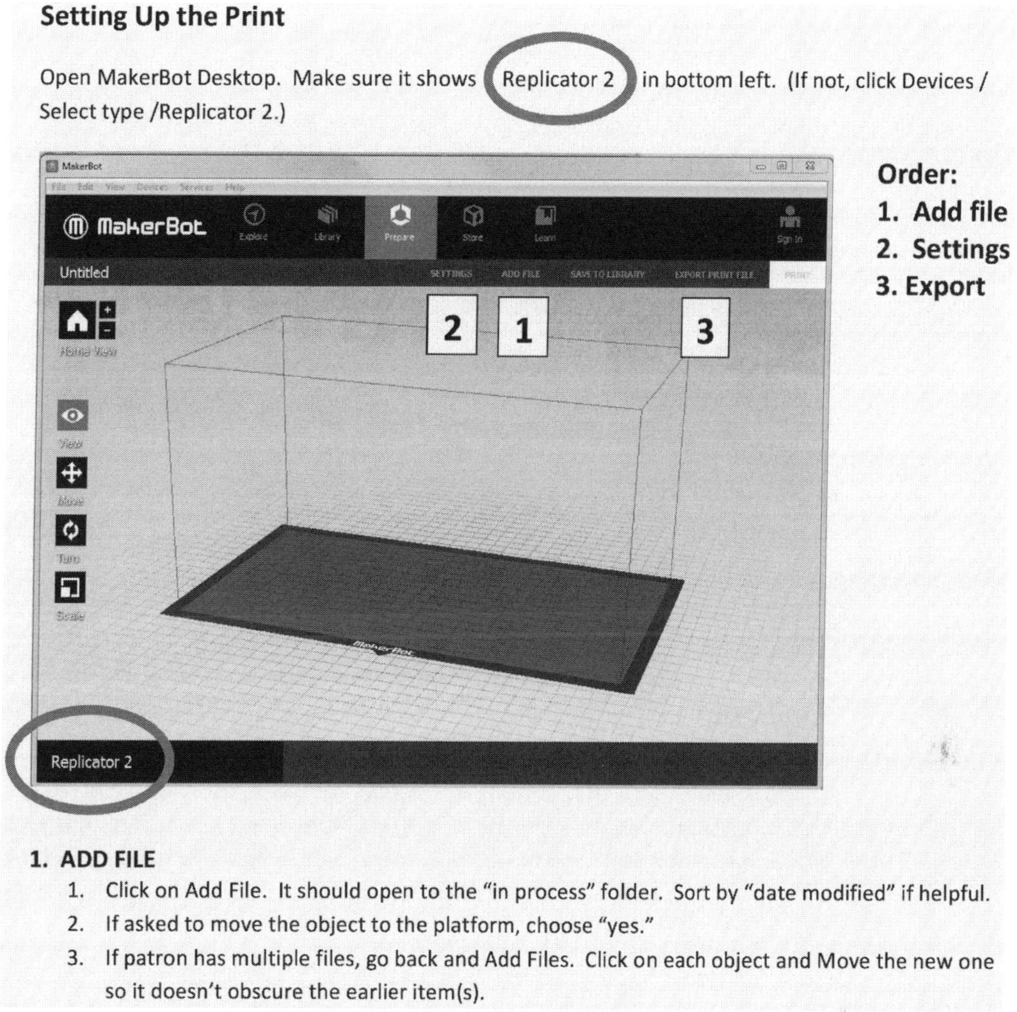

Setting Up the Print

Open MakerBot Desktop. Make sure it shows ⬭Replicator 2⬭ in bottom left. (If not, click Devices / Select type /Replicator 2.)

Order:
1. Add file
2. Settings
3. Export

1. ADD FILE

1. Click on Add File. It should open to the "in process" folder. Sort by "date modified" if helpful.
2. If asked to move the object to the platform, choose "yes."
3. If patron has multiple files, go back and Add Files. Click on each object and Move the new one so it doesn't obscure the earlier item(s).

Figure 9.3.   Marked-Up Excerpt from Staff Training Manual

patron-initiated service. For example, users are informed that "a 20% fill and random line fill pattern is a good starting point for the first print."

Whether you offer modeling services or not, staff must understand the basics of recognizing a bad model and referring appropriately. Again, the relationship between degree of difficulty of skills and staff acquisition of those skills is likely to resemble a pyramid, as shown in figure 9.4.

Recognizing bad models is a skill that will be developed over time. Training in the no-brainer aspects of recognizing a bad model is required for all staff who accept or

---

BEST PRACTICE EXAMPLES FOR ACCEPTING AND PROCESSING JOBS

- Every job: "lay flat" or "lay on plate" to reduce need for supports.
- Every job: rotate plate to examine underside of model.
- Filament choice: use white when user requires strength.
- Skinny pieces and sharp corners: add "pacmen" to extend the size of the raft.

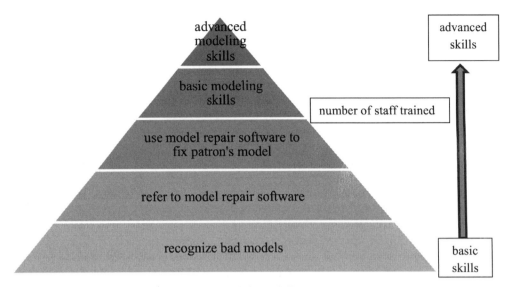

Figure 9.4. Pyramid of Expertise for Modeling Skills

process jobs. A "bad model" is likely to look quite different from a "good model" in your printer software (e.g., black instead of white, patterned instead of solid) and you should include an example, such as the one illustrated in figure 9.5, in your training documentation and your quick-consult manual as well as demonstrate the differences during a staff training workshop. Advanced "bad model" recognition is a skill that may only be acquired by the staff experts and may not need to be included in the general training. See chapter 4 for more details on what constitutes a good model and chapter 10 for additional examples of pitfalls during the file submission stage.

Figure 9.5 illustrates a "bad" model versus a "good" model. The view on the left should immediately trigger a "bad model" reaction because it is unevenly shaded, and black parts are immediately suspicious in the 3D printer software used in this example. The view on the right shows the same file after it was run through a model repair software (as described in chapter 4), and it now displays in appropriate shades of white.

Likewise, all staff who accept jobs should know the fundamentals of directing patrons toward fixes for problem models. The staff training workshop and manual should include your recommendations for model repair software and recommend that the patron should re-model an object that cannot be corrected by the printer software.

Figure 9.5. Bad Model versus Good Model

Payment

If you charge for jobs, selected staff must know how to calculate the total and accept charges. Your institution may require specific training of staff who handle cash- or card-based transactions, which may include awareness or procedures for protecting the privacy of patrons and their payment cards along with specific cash handling procedures. Those on your staff who do not accept payments (such as volunteers or student assistants) will not need to be fully trained, but they might still benefit from a general understanding of how the payment works, such as whether you accept cash or what sort of debit or credit cards are accepted.

Processing Jobs

Once a patron's model is accepted (and paid, if relevant), the job will need to be processed or "sliced" to convert the model into a set of precise "G-code" instructions, layer by layer, for the printer to interpret. This process can be both boring and time consuming, and it requires considerable computing power for large jobs. Your computers may dictate whether the processing function is to be handled with the model acceptance function or in a separate phase. In either case, staff who may generally perform only one function, for example, accepting or processing, should be adequately versed in the other function because they are highly linked and they might be performed iteratively (see chapter 10) until a patron is satisfied.

Basic Printer Operation

With a staff-mediated service, production includes removing jobs from the build plate, changing out filament when needed, and loading the next job. Depending on your volume of requests and number of printers, you may find it helpful to assign one or more staff to manage the overall production process at specified hours, which is likely to include making sure that the production keeps rolling and prioritizing jobs according to your methods.

Distill the steps from the user manual into a quick-consult version. The basic steps of checking and changing filament, loading the job, and removing a job from the build plate are quick to learn.

Queue Management

As the print requests roll in, you may find it helpful to assign one or more staff to manage the queue. These staff must be trained in your prioritization levels if you follow anything other than a "next in line" order (see discussion in chapter 6). The queue may benefit from adjustments throughout the day to keep the jobs printing efficiently, and your management procedures are likely to shift as your service matures. Any finessing of the queue management will need to be updated in your training documentation.

Notification, Pickup, and Closure

One or more staff members may be assigned to notify patrons when jobs are complete. All frontline staff will need to know how and where to check that a user's job is finished and

FAYETTEVILLE (NY) FREE LIBRARY'S POINT-OF-USE BOOKMARK

[Front Side of Bookmark]
Makerbot Instructions

1. Download a .STL 3D model from www.thingiverse.com, or create your .STL or .OBJ 3D model using 3D design software (Tinkercad, Sketchup, Solidworks, Rhino are all available in FFL Fab Lab).
2. Make sure Makerbot is on. If not, turn on with toggle on back of the machine.
3. Make sure the machine's USB cord is plugged into the computer you'll be using.
4. Download/open the .STL or .OBJ file you want to print. It should automatically open in the Makerware software. Move, rotate, and scale the object as desired.
5. To print, click "Make."
6. Adjust settings if desired. See the back of this bookmark for info on settings. Check the "print preview" box to see how long the print will take.
7. Select: "Make it!" when ready to print.
8. The machine will start printing—keep your hands outside of the machine.
9. Remove your piece. With the Replicator 2 machines, use a metal spatula if stuck. See staff for help.

[Back Side of Bookmark]
Makerbot Troubleshooting Guide

1. I clicked MAKE . . .what are all these settings?

Resolution controls how far the build plate moves down each pass. Higher resolution takes more time but better quality prints.

Rafts can be used for models that have a small surface area on the build plate.

Supports can be used for models that have many unsupported horizontal features.

Tip Click "USE DEFAULTS" to reset the print settings after a previous print.

2. My print is not sticking to the build plate.

Try to reprint using a RAFT.

Remove some parts (plate should be less than 75% full).

If this still doesn't work—try printing without painter's tape.

If this still doesn't work—run the "Level Build Plate" function under "Utilities." See a staff member for help.

what procedures to follow when giving items to patrons. Include instructions for closing or completing the job on any internal documentation.

An example of a point-of-use patron aid is found on a bookmark developed by the Fayetteville (NY) Free Library (see the textbox) and available at http://www.fflib.org/images/pdfs/MakerbotInstructions.pdf. Instructions don't need to be lengthy to be effective!

An example of a point-of-use staff aid is a large sign placed next to the 3D computer at the University of Florida's Marston Science Library. It outlines the basic steps and the order in which they should be performed.

Calling In the Experts

As with any library service, staff should know when and how to call in the experts. With 3D printing services, staff should know if your experts are available during set hours, on call, or by appointment.

Gently encourage your staff to build up their expertise by continuing to shadow the expert(s), rather than referring and running. This way, the burden of patron interaction and production management will not fall on one or very few. Recognize that the expert's knowledge base is also growing rapidly in the early months of service, so additional training as well as shadowing may reap benefits for your entire staff and thus for your patrons.

Troubleshooting

If staff are routinely observing the printers, they will quickly learn to recognize problems. They should be trained to stop the printer or call for help:

- When filament is not extruding from FFF printers. The printer may continue to "air print" until it thinks it has completed the job.
- When filament is globbing around the extruder nozzle instead of laying on the build plate.
- When "spaghetti" is being extruded.

With patron-mediated services, basic troubleshooting should be covered in training workshops or in documentation left by the printers. Make sure patrons know when to

⚙ University of Florida

Rafts or layers too tight or too loose?

Test line of filament squirting out too tightly?

- Level the build plate.
  - Adjust the distance between plate and extruder.
  - Make sure no filament is hanging from extruder.
  - Don't bother to adjust screws first. It's a time-waster.
  - Use something slightly sturdier than thin paper—such as lightweight card stock.

Filament jams?

- Get printer cool enough to work with, but filament may need to be heated for removal.
- Disassemble the working parts:
  - Remove fan parts.
  - Push back the motor.
  - Use pliers to grab filament piece.
  - Squeeze compressed air all around.
  - Replace the motor, reassemble the fan parts (label inside).

Filament won't unload?

- Press load, push filament down, then unload.

Air printing?

- Did you run out of filament?
- Did the filament jam on the spool?

Motor loose?

- Be sure to check the screws on the bottom of the motor.

Proportions off / holes not round?

- Tighten belts.

Belts loose?

- Follow instructions to move pulleys.
- Print your own clips.

Printer squeaking?

- Grease it! Should be done routinely, but if not, grease upon squeaking.

Temperature not holding?

- Replace ceramic insulation tape (refer to instructions).
- Check the fan—lettering faces IN.

Blue tape bubbles and wrinkles? We think the blue tape stretches when heated.

- Try to stretch the blue tape when applying, whether in sheet or roll form.
- Be sure to spatula down the tape between jobs.

Want to change filament? Running out, or want to change color?

- Press the M and PAUSE job (do not cancel).
- Unload and load; resume job.

notify staff of problems. More details about troubleshooting problems beyond the basic level are covered in chapter 10.

The Fayetteville [NY] Free Library reassures patrons via its Makerspace FAQ (2014) that "our IT team of 2 staff members take care of troubleshooting, repair and maintenance. We also purchase care plans that provide support and troubleshooting from equipment manufacturers when available." This public note communicates that the library has a support structure in place while offering a subtle reminder that machines can and probably will fail.

The University of Florida Libraries developed an FAQ sheet of basic troubleshooting for MakerBot Replicator 2 printers (see the textbox). Some of the repairs are considered basic while others require expert-level surgery, but the symptoms have been listed on one sheet to enable staff to define and report the problem efficiently even if they haven't been trained in all of the repair functions.

Maintenance and Repair

Routine maintenance will vary by printer and by hours of use, but the basics can be performed by many staff. With patron self-service, maintenance tasks will still need to be performed by staff or by a designated volunteer.

Examples of routine maintenance include watching and listening. If you use protective tape on the build plate, it must be examined regularly for tears or bubbles and changed out when needed. Moving parts may need to be lubricated as specified by the manufacturer. Rattling sounds may indicate loose bits of filament or other debris in the line. Loose parts may indicate a need to tighten screws.

Repairs for any model of printer will vary in degree of difficulty and access to the problem. Scour your user manual, the company's website, and user boards to identify the common problems and the level of expertise needed to solve them. See chapter 10 for additional guidance for troubleshooting printer problems.

Identify the moderately challenging tasks that might be manageable by several staff, and the expert repairs that should only be attempted by staff "surgeons." See figure 9.6 for the "pyramid of expertise" for repair functions. Your staff training manual should include illustrations of the repair, or links to relevant websites and videos. Save the complex repair tasks for the staff who are best suited for acquiring the skills and dismantling your equipment. You may be surprised to learn how much more quickly your surgeons can complete a repair after their first attempt. Understanding how the parts interplay and how to get to which set of screws is a major element in the battle. Take advantage of occasions when the printer is not in use to let several staff disassemble, examine or clean, and reassemble the printer. The familiarity will pay off during crises.

Some repairs cannot be managed in-house, either because the printer is so sealed that you cannot get to the problem part or because a part needs to be replaced. Make sure your staff understand which assemblies they should not attempt to open. All staff should know to whom and how to report malfunctions, but you are likely to prefer that only one staff member be designated to contact the manufacturer when needed.

⊚ Training Documentation

Consider developing several types of staff aids to provide efficient assistance at each point in the process. As the text above indicates, your staff will benefit from a range of documentation at several levels of detail, preferably accessible at service points, in staff areas, and online.

Any documentation created for the public can also serve as quick-consult or training tools for staff. Remind staff, especially those who are not involved with the service, that their basic questions are likely to be answered through the documents you create for your patrons.

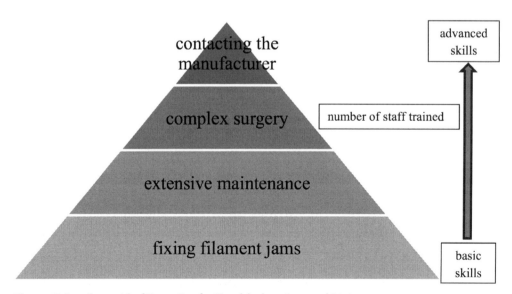

Figure 9.6. Pyramid of Expertise for Troubleshooting and Maintenance

TRAINING DOCUMENTATION SUGGESTIONS

Printed materials

- Public handouts. In addition to benefiting your patrons, handouts can serve as a refresher or starter for unfamiliar staff.
- Point-of-use signage and cards. Use sheet protectors or make large signs as appropriate. Place at the critical staffing points:
  - Public services desks
  - Desks where jobs are accepted
  - Desks where jobs are processed
  - At the printer
- Step-by-step training instructions.
- Printer manual.

Online guides

- Duplicate any information contained in printed materials (above) that is appropriate for patrons on your website or in public guides.
- Store longer versions of the items above that are not easily consulted in print format.

Internal website

- Step-by-step training instructions.
- Printer manual. Save and store a copy on your staff website.
- Links to online help, including the manufacturer's website. Seek out videos for basic and maintenance functions. Again, save and store copies to ensure continued access.

Public website

- FAQ about the service, including your policies.
- Guidelines for patrons, whether you offer staff-mediated or patron-initiated service.
- Photos of your samples, patrons' successful prints (with their permission), and examples of setting options or failures.
- Links to videos for basic functions.

Social media

- FAQs, how-tos, and best practices for 3D printing tend to be lengthy. Your patrons may be best served via social media with links to documentation posted on your website.

Download and save documentation you find on the web. Don't rely solely on links because they may disappear, especially when your printer model is upgraded or replaced or if the hints are stored on an individual's site that disappears when that person moves or is reassigned.

As you develop staff training materials, consider including lots of illustrations instead of text, interspersing workflow with operational steps. This strategy will enable your staff to find and implement the needed step quickly.

The size of your staff and your culture of using print or online training and support materials will guide your choices. The textbox suggestions cover all the bases, and may be overkill for smaller libraries.

⊚ Key Points

Staff training for a 3D printing service is a considerable undertaking. Tips to make the process work smoothly include:

- Provide training early for staff to increase expertise and buy-in for a new service.
- Identify the skill sets of your staff, and initially assign tasks to those with the greatest comfort level.
- Create appropriate documentation for each service element.
- Distill a staff manual from the products' user manual, website, and user boards. Wherever possible, use images rather than text for quick and easy consultation. Augment with your workflow for easy training and consultation.
- Develop training materials to suit both staff training and patron training, if appropriate.

Brace yourself to learn about the typical problems associated with 3D printers and with running a 3D printing service in the next chapter.

⊚ References

Fayetteville [NY] Free Library. 2013. "How To: MakerBot Replicator 2 Using MakerWare from the FFL." YouTube video, 5:48. Posted by Fay Free. September 4. https://www.youtube.com/watch?v=8Rh_xbIuRpk&index=5&list=PL9hcubxiao_2D_m1teBx22G1BJtkpjyTw.

———. 2014. "Makerspace FAQs: 21 Questions about FFL Makerspaces." Fayetteville Free Library. http://www.fflib.org/make/makerspace-faqs.

———. 2015. "MOJO 3D Printer Certification." YouTube video, 13:42. Posted by Fay Free. March 25. https://www.youtube.com/watch?v=UAWP81n_s0E&index=2&list=PL9hcubxiao_2D_m1teBx22G1BJtkpjyTw.

Gloucester County [NJ] Library System. 2015. "Guides/Forms." Gloucester County Library Commission. Accessed August 10. http://www.gcls.org/makerstudio/guidesforms.

Scalfani, Vincent. 2015. "The University of Alabama Libraries 3D Studio Standard Operating Procedures." Version 6.0, July. http://guides.lib.ua.edu/ld.php?content_id=5524864.

Dealing with Difficulties

Failures Will Happen

Prints fail for all kinds of reasons: high humidity, bubbles in filament, uneven powder spread, lens occlusion, lens misalignment, clogged nozzle, xy drift, bed misalignment, filament grinding, insufficient material, dull razors, curling, over/under extrusion/exposure/deposition, drooping. . . . Of course, parts with improper designs may also fail post-print, but that's the whole point of rapid prototyping! (Benster, 2014)

ALL 3D PRINTERS WILL, AT SOME POINT, EXPERIENCE some type of trouble such as a filament jam, loose belt, or model that won't adhere to the build plate. These problems are not something to fear but rather, with knowledge and preparation, they can become just a small roadblock to successfully printing again. It is important to remember that 3D technology is still in its relative infancy, especially compared to other types of library or office technology, and improvements are constantly being made with firmware, components, filament, and software.

There are several steps you can take, even before your printer arrives, to become prepared to tackle these challenges. The first step, and perhaps the most important, is to become fully acquainted with your printer. Take the time to get acquainted with your printer by reading through the manual, watching videos produced by the company and other users, and subscribing to manufacturer forums.

The next step is to develop your troubleshooting skill set. Become familiar with your tool set, and identify which screwdrivers and hex wrenches are needed for key components. Take your printer apart, as far as you can, always keeping in mind the limitations of the printer manufacturer warranty. Some printers are designed to be disassembled to have components replaced while other printers are closed and not suitable for dismantling by the user. By disassembling and reassembling your printer during a relatively stress-free practice session, your staff will learn what they can do on their own and will develop their ability to disassemble quickly and effectively during disasters.

Types of Printing Disasters

The frequency of problems will vary by printer type, but you are likely to encounter several of them. Table 10.1 shows the recorded (but underreported) rate of problems experienced by the University of Florida's Marston Science Library in the first sixteen months of service. Although reprints still must be run at a higher rate than desired, the staff have learned to detect and correct printer problems earlier in the process, reduced the rate of setup errors, and improved in the rate of detecting bad models before jobs are printed. This is demonstrated by the reduction in total problems by 2015.

Below are several categories of problems you will encounter. For more details and different perspectives, consult the website of 3dprintingforbeginners.com (2015).

Filament Issues

Problems with the filament feeding in smoothly to the 3D printer will likely be your most common issue to troubleshoot. There are many potential troublesome spots as the filament travels from the spool, through the guide tube, and then is heated within the extruder head, extruding down through the nozzle. A spool can run out of filament or tangle in the middle of a job. Filament can break or jam at any point along its travel path. A tiny bit of leftover filament in the extruder or guide tube can cause a clog. Other triggers for a clogged nozzle include when the filament fluctuates in diameter or is affected by humidity in the air. All of these scenarios can cause filament to stop flowing out of the nozzle, termed "air printing," and lead to a job failure. Figure 10.1 shows an example

Table 10.1. Types and Frequencies of Problems Experienced

TIME PERIOD	PRINTER OR FILAMENT PROBLEM	SETUP PROBLEM	MODEL PROBLEM	TOTAL PROBLEMS (RECORDED)	TOTAL PRINT JOBS
April–Aug. 2014	7 4%	4 2%	7 4%	18 9%	189
Sept.–Dec. 2014	37 8%	15 3%	23 5%	75 17%	445
Jan.–April 2015	19 3%	6 1%	3 0.5%	28 5%	581
May–July 2015	4 2%	1 0.4%	4 2%	9 4%	227

Figure 10.1. Megalodon Tooth

of a Megalodon tooth that failed because the filament tangled and stopped extruding. The wispy top layer exposes that it stopped printing because of a filament problem and not a model error.

Most 3D printers do not have a sensor that stops the printer if filament becomes clogged, and thus the printer continues to move as programmed until the user notices and cancels the print job. There is no way to recover the 3D model after air printing, and you will have to remove the failure and start over with a cleaned build plate, losing time and wasting material. This is one of the top reasons that 3D printers should be placed within easy view of staff and patrons in order to catch these unrecoverable errors.

At the University of Florida, a filament failure is considered the fault of the printer and the model is always reprinted at no charge to the patron. Different types and colors of filament can exhibit different materials properties, and some delicate jobs may print successfully if you try a different (e.g., less brittle) filament.

BEST PRACTICES FOR REDUCING FILAMENT PROBLEMS

- Check the level and color of filament before beginning each job.
- Check the filament frequently for brittle properties.
- Stored unused filament in plastic bags with desiccant packs to reduce the effect of humidity.
- When removing the spool, catch and secure the loose end of filament to the spool for storage. This prevents the loose end from diving under other loops and tangling during a future print job. For additional guidance on how to untangle a spool of filament, follow ToyBuilder Labs' tutorial and video (ToyBuilderlabs.com, 2015).

Printer Setup

Failures will result if the printer is not set up correctly. Parts need to be checked frequently:

- The build plate must be level and at the correct distance from the extruder nozzle. The manual will include instructions for adjusting the distance and levelness of the plate and should be frequently checked. Test and adjust these functions when you set up the printer, but also retest if you think items are not printing accurately or after you have moved the printer.
- Due to printer vibrations or relocating a printer, screws may loosen and lower the accuracy of the printer head. Figure 10.2 shows a model printed on a 3D printer with a loose x-axis belt. As the print head traveled in a circular motion, the slack belt affected its accuracy, resulting in a gap or crack in the model. Loose components may be audible or even damage other components if allowed to vibrate unchecked. Check periodically for components that wiggle and test belt tension to ensure that the printer is structurally sound.
- You may need to calibrate your printer, especially before its first use. In a properly calibrated printer, the extruder will move along the x, y, and z axes at a known rate and the printer will extrude an exact amount of filament. When you print circles or holes that appear oval in a finished print, or if the model dimensions are obviously incorrect, you should immediately run a calibration test. To check your printer's calibration, print out a calibration sample as recommended by the manufacturer or from Thingiverse.com (e.g., http://www.thingiverse.com/thing:24238). Measure all dimensions of the test print to check the calibration, and then adjust as necessary according to the printer instructions.

Figure 10.2. Model Printed with Loose X-Axis Belt

Warping

Depending on the size, geometry, and length of printing time, you may discover that some models warp while printing (see figure 10.3 for an example). This is due to uneven cooling of the model layers when lower layers cool more rapidly and shrink compared to the warmer upper layers. This warping effect is especially noticeable on large, flat models (such as a box) that require a long print time. Some warping is cosmetic and may be tolerated by the patron, whereas other warping can be so severe as to lead to print failure, especially if the model curls up sufficiently to interfere with the extruder nozzle. Warping is more likely with ABS filament rather than PLA due to different shrinkage rates (Fab-baloo.com, 2014) and may be exacerbated by temperature changes (such as wind gusts) or a dirty print bed.

You can take several steps to mitigate the problem of warping. Recommended best practices are included in the textbox.

For additional suggestions for preventing warping, see the blog posting on the MakerBot website (MakerBot Industries LLC, 2011), which applies to printers produced by other manufacturers as well.

Models Detach from Plate

Warping can cause the edges of models to pull up from the plate, but you may also encounter entire models popping off the build plate during printing. Once this occurs, there is no way to recover the job and you must cancel the print, clean the bed, and restart. The only exception is if you are printing multiple pieces and some pieces are still attached to the bed and printing successfully. Figure 10.4 shows a print composed of three models on the same build plate where one model detached, but the other two models continued to print successfully. Figure 10.5 shows a cylindrical model that detached but the printer continued to print, producing "spaghetti" filament. To avoid models detaching prematurely, make sure that the print bed is clean and level before each print job. Some models may need a raft or brim to stay attached if only a small portion of the model would be attached to the build plate.

BEST PRACTICES FOR REDUCING WARPING

- Ensure that the print bed's surface is clean and level so that the first layers adhere as much as possible to the bed.
- Experiment with different print bed surfaces to find the type that is the "stickiest."
- Add "pacmen," wafer-thin three-quarter circles, to the corners of large models to increase the footprint and adhesion to the plate. Search for "pac-man" on Thingiverse (http://www.thingiverse.com) to see examples of these helper discs that can be resized as appropriate.
- Minimize drafts around the printer by adding an enclosure or by controlling the airflow.
- Print slowly to allow the layers ample time to cool.
- Rotate the orientation of the model, if possible, so that the model design will be least affected by any warping.

Figure 10.3. Warping

Extruder Temperature

The best print jobs result when the filament is heated to the appropriate extrusion temperature. If you see excessive drooping or if the filament is extruding too solidly, experiment with altering the temperature at five-degree intervals when processing the job. You may find that different filaments, and even different colors of the same filament, extrude better at slightly different temperatures.

Also, watch the printer thermostat closely to see if the extruder is maintaining the desired temperature range during printing. If the temperature fluctuates during printing more than one or two degrees, consult the manual or manufacturer for recommendations. The solution may range from replacing the extruder to replacing insulation around the extruder head.

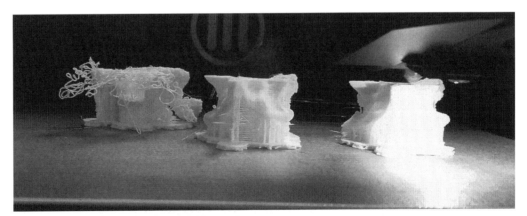
Figure 10.4. One Item Detached from Plate

Figure 10.5. "Spaghetti" Happens When Detached Item Is Not Level

Power Outages

If your library or your 3D printer suffers a power outage during a print job, the print job will likely be unrecoverable and will require restarting. Most printers will not maintain data about the point reached in the job when the power was lost. As reluctant as you may be to pull the plug during a test job, it might be useful to know how your printer reacts: does it shut off or power back on, and does it remember its place? If your library is in an area with frequent power outages, consider investing in an emergency uninterrupted power supply (UPS).

Failure Due to Bad Models

Some jobs fail because the model is imperfect. Models with errors may still print correctly, or portions may be missing or incomplete, or the entire model may not print at all. As outlined in chapter 4, learning to recognize bad models during the processing stage will save your printer from wasting time and filament by generating flawed output. Ideally, staff will identify a problematic model before printing and encourage the patron to correct the model or locate a new one to print. Decide whether the library will reprint at no charge to a patron for a bad model or if the financial onus is on the patron (see chapter 7 for full policy discussion).

⊚ Patron-Initiated versus Staff-Mediated Troubleshooting

The level of responsibility for solving failures will vary significantly between patron-initiated printing services and staff-mediated service. With a staff-mediated service, you can control the level of training and responsibility expected of each staff member. In this model, the staff are also familiar with which problems are common to your particular 3D printers. Make sure your experts are available during peak printing times and all staff know the protocol for troubleshooting and for contacting the experts.

If you have a patron-initiated service, make sure that your troubleshooting staff are available during printing hours to assist as needed. Establish what level of repair you will permit your patrons to perform. Be explicit about what they may or may not do, such as whether they may remove filament or what pieces they may unscrew. Prepare instruction and signage that indicates when and how patrons should contact staff when problems arise.

Extended jobs do not necessarily indicate a printing problem, but they may indeed cause service problems. "If your print goes over your reserved time limit and you did not extend your reservation, your print will be canceled" (University of Michigan Library, 2015). Determine early on your policy for what patrons should do if a print job is continuing past a reservation time. This may require you to adopt conservative time restrictions since some 3D processing software drastically underestimates the expected print lengths.

⊚ Managing Patrons' Expectations

You will find patrons who know absolutely nothing about 3D printing, some who know a great deal, and some who strongly believe they know more than you do. Figuring out the limits of a patron's expertise can be challenging, and staff should be encouraged to assume a lesser knowledge base until proven otherwise. Even the experts among staff and patrons will make mistakes and invalid assumptions, and a certain rate of failure is inevitable. It is important to remember that 3D printers vary significantly and, even if a patron has experience with 3D printing, it does not mean that the patron has expertise using your library's 3D printers.

Patrons' Expectations

Patrons' expectations for turnaround time and print quality will vary a great deal, but you may be surprised to see how many of them have lower expectations than your staff have. Some patrons will expect their jobs to print immediately, but many will be fine with a few days or a week. Following are some key expectations to anticipate and address.

Immediacy

Patrons may expect a job to be printed upon submission, as they may not anticipate that other patrons' jobs may be waiting in the queue. Other patrons assume that the time to process a job is instantaneous, not understanding that variables must be set and models must be sliced and saved before they can be sent to the printer. Make sure that your marketing materials, whether on your website or through social media or in-library, state your typical length of the submission process and the queue, and are adjusted appropriately during times of heavy submissions.

Speed/Duration

New users may assume that a 3D print will take no longer than a paper print or photocopy. Some are quite dismayed to learn that, since 3D prints are built up layer by layer, they may take hours to print. Again, try to manage expectations through marketing to reduce the disappointment upon submission. Locating the printers in a space viewable to the public can help temper expectations since they can see the print speed themselves. Sample pieces with time estimates are also helpful for patrons who just want to know how long it takes to print out a model.

Accuracy and Perfection

Patrons who are creating engineered parts that must fit into something may have high or unrealistic expectations of your 3D printer's accuracy. In most other situations, accuracy will be neither expected nor observed.

Assistance with Modeling

As stated in other chapters, you must decide what level of modeling assistance your staff will provide and advertise accordingly. Be honest with your patrons about your staff's expertise and computer limitations. Develop a list of experts who are willing to assist patrons with their modeling questions and problems.

BEST PRACTICES FOR MANAGING EXPECTATIONS OF ACCURACY

- Don't promise accuracy or perfection!
- Understand the limits of your printer. The manual should address the likelihood of accuracy (within a percent) and the typical rate of shrinkage expected with various filaments.
- Measure finished pieces, and compare the measurements against the original model. Measure on a regular basis (e.g., monthly) and after the printer has been moved.
- Print calibration samples (found on Thingiverse.com or recommended by your manufacturer) and adjust your belts as needed and described in your printer manual.
- Provide recommendations for other 3D printing services with better 3D printers (nearby and online) for patrons that demand a high level of accuracy and precision for their model.

Problem Patrons

As with all library services, you will encounter problem patrons. Prepare your staff to cope by using some of the following strategies.

Don't Know What They Want

Some patrons will simply want to 3D print "something" and will not easily grasp the idea that they must first select a model. Patience on the part of staff is required. Website guides, flyers, and 3D printed sample pieces can help guide the patron toward a model suitable for printing. Consider selecting a well-tested popular model to suggest to patrons who are seeking first-time or generic gift models.

Keep Changing Their Minds

You will encounter the type of patron who submits a job while insisting that all the variables (quantity, size, infill, filament color, etc.) are set. Then, during the processing stage, the patron changes one or more of the variables, requiring that the job be reprocessed. If you offer staff-mediated service, learn to recognize these patrons and notify all involved staff to clarify these patrons' initial settings and to double-check each time that they are committed to the variables requested.

When Rapid Prototyping Becomes Repetitive

3D printers allow users to benefit from **rapid prototyping**—a technical term that refers to quickly drafting a model via computer-aided design and producing a working copy. The next stage is to assess the copy, redraft the model, produce another draft, and repeat. However, some patrons will get carried away with the iterative process, continuing to submit draft models that may or may not solve the design problem. Staff will need coping strategies when they feel that a patron is crossing the line of continued bad design, staff patience, and dominating the printer queue. The textbox summarizes the best practices for working with patrons who seem to keep looping through the same request.

BEST PRACTICES FOR MANAGING REPETITIVE SUBMISSIONS

- Work with the patrons who request one print of small and quick items. Encourage them to select the same type of filament as another print job so that their model can be quickly printed between longer jobs.
- Firmly state when you recognize that new versions of a model still have errors, such as when a side is still too thin or if the model is likely to fail while printing.
- Call in the colleague who has previously assisted the patron with the patron's earlier model for assistance in recognizing the printability of subsequent models.

⚙ Examples of Tips for Users

The textboxes present two examples of best practice documents from the McGoogan Library of Medicine at the University of Nebraska and the Marston Science Library at the University of Florida. These checklists are intended to illustrate the granularity of troubleshooting advice given at two locations. The specific guidance will not be relevant to all libraries or to all printers.

PATRON-MANAGED SERVICE EXAMPLE

At the McGoogan Library of Medicine (2015) at the University of Nebraska, patrons use the printers themselves while staff act as facilitators. The following instructions for patrons are provided on the library's website.

3D Printing Tips and Tricks For MakerBot Replicator 5th Generation

- The print tray should be clean of debris and excess filament before printing.
- The tray should be covered in MakerBot Build Plate Tape or blue painters tape.
- The printer has better X-Y accuracy than Z-accuracy, so if you want accurate holes, pillars, cylinders, screw threads, etc. in your objects, arrange them so these objects rise up vertically instead of laying them down horizontally.

Homing Error: If the printer stops your job before printing because of a homing error, reprint the job. If it continues to say homing error, or if it says homing error and stops your job midprint, contact the Systems Dept.

Filament Not Extruding: If the filament does not extrude, try unloading filament and reloading. Allow the printer to rewind the filament entirely before pulling filament out. If filament does not pull out ask for help or contact the Systems Dept.

STAFF-MANAGED SERVICE EXAMPLE

The Marston Science Library, University of Florida (2015) created an internal document for staff to consult when problems occur with MakerBot printers. Since staff manage the printers, this checklist assumes a prior level of training in using the printers. It is written in a telegraphic format for quicker consultation in situations where a reminder, not an explanation, is needed.

3D Printing for Dummies—Checklist

1. Always **check the SD card** before removal to make sure the file has saved. Refresh the card contents if needed. Change the filename if it still won't behave. Then close the SD window before removing card.
2. Always make sure each object is **ON PLATFORM** (move button) or **LAY FLAT** (turn button). If not, you may get (a) layers of support material between object and raft or (b) spaghetti if no supports were added.
3. Check **filament level** on spool (see "grams left" note). Use small pieces whenever possible.
4. **Size matters!** Especially if an object appears very small on the plate, ask the patron about expected size. Each big square is 1cm; each fine square is 2mm. Use the Scale button to see the sizing.
5. Not sure about **supports**? Process the job WITH supports and see what they look like. Assess degree of difficulty in removal. Try other positions of the object.
6. **Don't see the file on the SD card?**
   - Scroll up and down the card.
   - Take the SD card back to the computer and refresh it.
   - Sort the SD card by newest first.
   - Change the filename on the SD card (preferably shorter).
7. Why are you getting **spaghetti**?
   - Job may need supports or raft.
   - Object may need to be turned and reprocessed.
8. **Model advice**
   - Check **Print Preview** after exporting file and scroll through the layers.
     - Keep flipping the object if needed, export and print preview, and note grams/time to find the most efficient setup.
     - If the piece is functional, ask which side is most important to be clean or precise. Don't lay the job so supports are built on the cleanest side.
   - If support material looks too difficult to remove, our printers may not be suitable for the job.
   - Bad model? Recommend patron run it through Microsoft Model Repair Service, aka NetFabb.
   - Resizing needed? If extreme, best for patron to remodel and return.
9. **Objects/job**
   - Don't put too many objects on the plate at once. Jobs that take a long time tend to have raised edges.
   - Lots of small pieces? If tiny (≤4 cm), put them close enough to get a common raft. If pieces are >4cm, leave space in between for separate rafts to result in less curling.
   - If patron has several files, create several slips (especially if different colors) but then add up all the subtotals for grams and minutes and only charge them once.
10. **The patron is impatient**

> - If patron wants multiple copies, it's OK to process one and multiply the single cost to get the total grams and time for charging purposes. Then charge them once.
> - If lots of pieces in the job, explain that the process takes a lot of time. Offer to process the job and notify patron of estimated cost later via email, then patron can return to OK and pay. Then we'll start the job(s).

Key Points

Even the best 3D printers will fail sometimes; with experience and practice, your staff can learn to minimize the errors and to troubleshoot efficiently. Following are tips for minimizing problems before they occur.

- Test your printer before launching the service so you get an idea of what problems to expect.
- Become familiar with the printer manual and online videos for how to troubleshoot common errors.
- Develop procedures for managing failures, redos, and backed-up appointments.
- Develop strategies for handling your "problem patrons."

The next chapter will discuss some marketing strategies, outreach opportunities, and workshop options that you can establish at your library.

References

3Dprintingforbeginners.com. 2015. "11 Experts Share Their Top 3D Printing Beginner Tips." *Beginners Corner* (blog). May 6. http://3dprintingforbeginners.com/3d-printing-tips/.

Benster, Tyler. 2014. "What Might the Future of 3D Printing Look Like?" [guest post] *3DaGoGo Blog* (blog). July 2. http://blog.3dagogo.com/.

Fabbaloo.com. 2014. "Shrinkage: A Problem of 3D Measurement." *Daily News on 3D Printing*. Fabbaloo. May 8. http://www.fabbaloo.com/blog/2014/5/8/shrinkage-a-problem-of-3d-measurement.

MakerBot Industries LLC. 2011. "12 Ways to Fight Warping and Curling." *Blog: Research and Development*. June 23. http://www.makerbot.com/blog/2011/06/23/12-ways-to-fight-warping-and-curling.

Marston Science Library, University of Florida. 2015. "3D Printing for Dummies." Internal staff document. February 27.

McGoogan Library of Medicine. 2015. "3D Printing @ McGoogan Library: Tips and Tricks." McGoogan Library of Medicine. University of Nebraska Medical Center. Accessed August 11. http://unmc.libguides.com/content.php?pid=653308&sid=5410705.

ToyBuilderLabs.com. 2015. "Untangling the Filament Spool." ToyBuilderLabs. Accessed August 11. http://www.toybuilderlabs.com/blogs/news/13055029-untangling-the-filament-spool.

University of Michigan Library. 2015. "3D Printing—Walk Up Printing FAQ." UM3D Lab. Digital Media Commons. University of Michigan. http://um3d.dc.umich.edu/3d-printing-walkup-faq/.

CHAPTER 11

Outreach and Marketing

IN THIS CHAPTER

▷ Creating effective publicity and marketing

▷ Offering 3D printing workshops

▷ Taking 3D printers out of the library into the community

▷ Developing outreach activities

ONCE YOUR 3D SERVICE IS OPERATIONAL, it will be crucial to publicize this new service through outreach events and reaching out to new users through workshops and activities. You may choose to recruit volunteers, host workshops, and develop marketing strategies to sustain your service.

Publicity and Marketing

As soon as you begin developing your 3D service, start planning for how best to unveil and publicize your new printers and related services. Although 3D printing is still considered an emerging technology, publicizing it should follow the same strategy you already use for other more standard library services, including the support of a library marketing team or publicity expert if possible.

The first step is to advertise the forthcoming 3D printers. Patrons will likely care about the type of printers, the maximum print dimensions, any costs to print, what type of models they can print, and when the 3D service will open. Place signage near the future location of the printer and, once the printer is in testing, create a gallery of models to highlight the potential of 3D printing. You may be tempted to only display the successful prints, but also keep the failures since they can either provide warnings on how not to print or illustrate a key concept of printing. For example, a model that is missing its upper layers due to a filament jam will reveal to patrons the infill pattern and shells. Continue to expand and improve the models in the gallery after the service is operational, as it will serve as an inspiration to patrons new to 3D printing plus provide hands-on learning objects when giving workshops or impromptu presentations.

Once the service is fully open, you may still not want to advertise too widely in order for staff to have ample time to train on the printers and software. The first few months of a service can be considered a shakedown of your policies, procedures, and equipment, and it is wise not to encourage too much demand until the service is operating smoothly.

Depending on foot traffic, just placing the printers in public view may be enough marketing at the beginning since the light and movement will attract interested users, as discovered by the Fayetteville (NY) Free Library. "We also feature a 3D printer at our front desk, which draws many inquiries and causes many people who might not otherwise know about or be interested in our Lab to want to learn more" (Fayetteville [NY] Free Library, 2014). Install a sign next to the printers instructing patrons how to submit a model or make an appointment. Consider creating bookmarks or small flyers as well for patrons to take away with important details.

As your 3D service progresses, continue to analyze user statistics and assess the health of the service (as discussed in chapter 5). You may decide that the service is sustainable and well supported without the need for additional marketing. However, if the service and staff can handle more submissions or usage, there are many marketing venues to try to reach new patrons. Ideas for strategies and marketing opportunities include contacting:

- Hobby stores, who often have a community bulletin board for relevant announcements.
- Makerspaces and hackerspaces, to request that they publicize it to their members.
- Local middle and high school science teachers, especially those that have robotics or engineering programs. Many students participating in academic competitions, such as First Robotics (First Robotics Competition, 2013), are already using 3D modeling and printing to create specialized parts for their team submission. These students may also be ideal first users to help test your new printers and service, or even as assistants in workshops to introduce other students to 3D printing.
- Other nearby libraries, academic or public, who may also be operating their own 3D printers or holding complementary activities.
- Any local or regional maker fairs. Maker Faires, trademarked by *Make* magazine (Maker Faire, 2015) and other similar events, are festivals where makers of all types gather to show their inventions, learn new skills, and discover new innovations and technologies. These events are perfect opportunities to highlight your library's technology offerings and showcase your patron and staff creations.

In addition to the above publicity suggestions, social media offers another touchpoint for connecting to patrons and potential partners, plus it can be a source for helpful ideas. Most libraries already conduct social marketing using major services such as Facebook, Twitter, and Instagram (King, 2015). 3D printing is a natural fit to a library social media campaign, and an easy method to advertise the printers is to post photographs of new 3D models.

The University of Nebraska Medical Center's McGoogan Library of Medicine uses Pinterest to publicize their recent 3D models. "After printing: If you would like your work showcased on the McGoogan Library website, please send a .jpg, .gif, or.png image file to library@unmc.edu and we will post the photo to the Library's Pinterest site (http://www

.pinterest.com/unmc/mcgoogan-library-of-medicine-3d-printing/)" (McGoogan Library of Medicine, 2015).

Remember to obtain permission of the patron or model creator before posting on social media, since some users may not wish to have their model publicly displayed. To gather an idea of what types of photographs and text are already being used, search Twitter using the hashtag #3Dprinting.

⊚ Instruction, Workshops, and Tutorials

Workshops, especially introductory tutorials, are an effective method of reaching out to new users of 3D printers. Whether you offer patron-initiated printing or staff-mediated printing, your patrons will expect some level of instruction or workshops to prepare them for using your service. The amount and type of instruction will depend upon your staffing and expertise level of patrons. However, ideally the list will offer a mix of workshops for different proficiency levels and instruction in both printing and modeling. The format of the training does not have to be entirely in person since there is a wealth of online tutorials available, including videos, library guides, and step-by-step hands-on instructions. Table 11.1 provides a resource list of examples. As with other library services, you should plan for a mix of instruction types including formal scheduled workshops, online tutorials, and instruction on demand at the reference desk or at the 3D printers.

Table 11.1. Online Tutorials and Examples of Workshops

Autodesk's Tinkercad modeling software	Tinkercad offers interactive, step-by-step tutorials that walk new users through modeling basic objects to gain familiarity with software. http://www.tinkercad.com
St. Joseph County (IN) Public Library	Handout with instructions for searching Thingiverse, using Tinkercad, and preparing model to print using MakerBot's MakerBot Desktop. http://sjcpl.org/sites/default/files/Studio304-ThingiverseTinkercadMakerBotDesktopr.pdf
Boise State University	Brief descriptions of the range of workshops that have previously been offered. http://bit.ly/discoverbydoing
Brigham Young University	Holds classes in printing preparation, 123D design, Blender, Adobe. Prezis, class PPTs, and the files generated in workshops are posted online. http://guides.lib.byu.edu/c.php?g=216600&p=1429613
Colorado State University	Introductory training course outline at https://idea2product.net/getting-started/3d-printing-training/.
Gloucester County (NJ) Library System	Lists quickstart guides and handout tutorials in PDF format at http://www.gcls.org/makerstudio/guidesforms.
Keene (NH) Public Library	Description of weekly labs teaching Thingiverse, Tinkercad, and MakerWare (MakerBot's software to process models for printing). http://www.keenepubliclibrary.org/3Dprinter
Northbrook (IL) Public Library	Videos and online tutorials about 3D printing, modeling, and Thingiverse. http://www.northbrook.info/3d-printing
Riverdale Collegiate Institute	Online document about basics of 3D printing along with links to other tutorials and resources. http://goo.gl/M4e7li

Role of Volunteers

Volunteers who are knowledgeable about modeling and 3D printing can be a great asset to your training and workshop program. Ask patrons who seem to have the skill sets and communication skills if they would be willing to teach workshops. Those who are highly enthusiastic are likely to make acceptable trainers. To provide encouragement, suggest that the accomplishments will stand out on volunteers' résumés and offer to serve as a reference for their future employment applications. Another option, if you charge users, is to offer free 3D printing to your unpaid assistants. As with all volunteers, they may show up inconsistently or may not be available at your times of greatest need.

Developing Workshop Content

Rather than reinventing the teaching outline each time, develop a knowledge base of lesson plans, slides, and videos (on a shared network folder or website) for staff to select the examples and areas of emphasis for each new workshop. For each type of audience, select which components to address, how much time to spend on each component, and the balance of delivery from lecture/demo to video presentations to hands-on practice for each component.

Depending on the composition of the class participants, you will need to consider the level of complexity, broadness of material, and focus on library policies and workflow. Workshops composed of community members seeking an introduction to 3D printing will benefit from a general overview of how 3D printers operate by showing short videos and in-person demonstrations. Use case examples from many different fields (e.g., education, food, medicine, and art), and show how participants can download 3D models to print at the library. Do not go into detail about modeling or very technical aspects of 3D models or printing.

Like the general community, K–12 students will also benefit from a simple overview of 3D printing, but examples should focus on topics that will interest children, such as printing chocolates, ears, and Pokémon figurines. Live demonstrations of 3D printing will fascinate children, but take care that they stay safely away from hot components. Depending on the amount of time and access to computers or tablets, students can be taught how to create simple models using Tinkercad or the mobile app Tinkerplay (http://www.123dapp.com/tinkerplay).

College and university students will require a more complex explanation of 3D printing and the various print materials available. Focus on how to obtain a 3D model, including modeling software available to them at their institution and guidance on what makes a good 3D model. Hold modeling workshops separately unless you are familiar with the expertise of the students attending, since engineering, design, and architecture students will likely already have expertise with professional modeling software (such as SolidWorks or Inventor). Give a full explanation of your library 3D printing service and policies and be prepared for questions about using alternative materials and other nearby options for 3D printing.

Finally, workshops for library staff should start with a brief overview of the basics and uses of 3D printing to motivate why the library is offering such a service. Provide staff with information about where they can find or create their own models, to encourage development of in-house expertise, and you may wish to include a hands-on modeling workshop depending on computer availability. Plan to spend a large amount of time explaining the library policies and walking through the service workflow. Encourage staff to offer suggestions on how to improve the workflow and to pursue additional training in their area of interest such as operating and maintaining the printer, modeling, or 3D scanning.

With each new group, evaluate the success of each component of the presentation to refine the content and improve delivery. Table 11.2 details a framework that can be used for these sessions and the general components included in each type of session, based on various audience levels.

Ⓖ Examples of In-Library Workshops

The following case studies describe workshops that academic libraries have developed in collaboration with other groups to introduce students to 3D printing.

Case Study: A Collaboration between Education Faculty and the Library

Librarians at Southern New Hampshire University supplied these examples.

In September 2014, the Shapiro Library at Southern New Hampshire University (SNHU) moved into a new Library Learning Commons building and opened the Innovation Lab & Makerspace, which included two Makerbot Replicator 2X 3D printers. Mid-fall semester, a faculty member from the School of Education approached library staff about using 3D printing in her Elementary Science Methods class. During this first semester, the faculty member brought students to the Makerspace for a casual demonstration of 3D printing and a discussion about possible uses of 3D printing in early education classes. The faculty member indicated to library staff that this demonstration and discussion was important and that a more formal application of 3D printing in class curriculum could be valuable.

The following semester, the faculty member required her students to attend one of the Library's "Introduction to the Makerspace & 3D Printing" workshops, where they learned how to run the Makerbot Replicator 2X 3D printer. Students then came back to the Makerspace after selecting a 3D model freely available online which they felt could be used to teach a science or technology concept to early education students. Some of

Table 11.2. Workshop Framework

| AUDIENCE | COMPONENTS | | | | | |
	TYPES OF PRINTERS	ABOUT 3D PRINTING	USES FOR 3D PRINTING	MODELING DEMO	HANDS-ON MODELING	LIBRARY SERVICE
Community	Lengthy/ videos	Lengthy/ videos	Lengthy/ videos	Brief/video	No	Brief
Library Staff	Moderate	Moderate	Moderate	Brief/ separate	Brief/ separate	Lengthy
University Students	Brief	Brief	Brief	Moderate/ separate	Separate	Lengthy
K–12 Students	Brief	Brief	Brief	Brief	Lengthy	Brief

the objects students chose to print included models of plant and animal cells, human organs and bones, interlocking gears, topographical maps, miniature bridges, models of the Earth, a frog dissection kit, and more. This demonstrated to students that 3D printing could be a mechanism for obtaining teaching tools inexpensively and on demand and helped them develop and teach with models, a requirement of the Next Generation Science Standards. Many students were excited about 3D printing and using 3D printed models in the classroom.

This experience taught library staff the necessity of offering flexible dates/times for workshops when a large number of students would need to fit the workshop into their varied schedules. Library staff also found that, ideally, workshops should be held during regularly scheduled class times to reduce the number of alternative/additional sessions being offered. (Harris and Cooper, 2015)

The Next Generation Science Standards mentioned above "include the critical thinking and communication skills that students need for postsecondary success and citizenship in a world fueled by innovations in science and technology" (Next Generation Science Standards, 2015). For further details about SNHU's 3D printing service, see the guide at http://libguides.snhu.edu/makerspace.

Case Study: Middle School 3D Printing Workshop

The University of Florida's Society of Women in Engineering (SWE) and Marston Science Library partnered in spring 2015 to introduce local seventh graders to 3D modeling and printing. Students were divided into groups of thirty students and rotated through 105-minute-long sessions held in a library computer lab. SWE students volunteered to assist with the workshop so that there was at least one volunteer for every three students.

Librarians first started with a fifteen-minute introduction to 3D printing, while a 3D printer printed a small model in the front of the room for students to observe. A short demonstration of Tinkercad followed, and then students were given a handout with step-by-step instructions to create a small name tag (handout available at http://guides.uflib.ufl.edu/ld.php?content_id=8095443). The example tag was a 1" × 3" thin rectangle with a name in raised lettering that would be later printed by the library and sent to the students. It was designed to teach the students how to create basic shapes, add text, modify model dimensions, and group components for exporting to an STL file.

Students proceeded to create accounts on Tinkercad and work through the instructions, notifying the librarians when they were finished with their model. The librarians saved each model using a filename derived by the student's last name and preferred filament color.

One problem encountered during the workshop was that some students were younger than thirteen years old, and so required a parent's permission for Tinkercad registration. Another issue was that students wanted to design name tags that were larger and more detailed than instructed. Not all tags were "printable" and some needed to be modified or rescaled by the librarians afterward to print successfully. The students also worked at varying speeds and some were still modeling at the end of the workshop while others had finished much earlier and needed further engagement. Librarians directed these students to search Thingiverse.com and to try to create additional models using Tinkercad (Buhler et al., 2015).

A 3D printer provides an excellent opportunity to reach beyond the library to engage community groups, promote STEAM (science, technology, engineering, arts, and mathematics) activities, and support service and charity initiatives. Participating in these activities can be incredibly rewarding but also may require significant staff time or a financial commitment.

Possible opportunities include giving talks, demonstrating 3D printing, and providing free printing of models for worthy causes. The Cleveland Public Library reports that "TechCentral is often requested to perform outreach 3D printing basic programs to schools, community groups, and even other libraries whose patrons are interested in this up-and-coming technology" (Urban Libraries Council, 2015). Groups that may be interested in being included in your 3D outreach efforts include:

- K–12 schools: Possible activities include giving demonstrations and workshops to students, allowing academic teams to use 3D printers, providing guidance and training to teachers if schools are considering acquiring printers. The University of Texas at Arlington states that "the FabLab's mission includes outreach beyond the university. We are currently developing activities for K–12 students to enrich their experience and engage young scholars through Creation eXploration and Innovation" (University of Texas, Arlington, 2015). The Cleveland (OH) Public Library's TechCentral Makerspace has "partnered with the Cleveland Metropolitan School District's MC2 STEM High school to provide internship opportunities for students within the MakerSpace" (Urban Libraries Council, 2015).
- Retirement communities: Many retirement communities host educational classes and workshops that welcome lectures about emerging technologies.
- Service organizations: There are many volunteers using 3D printing to develop devices and tools for special needs populations. For example, Enabling the Future (http://enablingthefuture.org/) and Limbitless (http://www.limbitless-solutions.org/) are designing and fitting patients with 3D printed prosthetic hands and limbs.

You are likely to be invited to provide demonstrations or workshops to groups that prefer to meet outside of the library. Whenever possible, attempt to schedule these activities in the library because taking the printer out of its queue and moving it (as well as necessary supplies) out of the library is not a trivial task.

If you must take the printer away from its home, use these guidelines to securely transport it and ensure successful printing posttravel:

- Use a wagon or dolly to carry it securely.
- Check ahead of time to ensure that there will be a sturdy, flat surface large enough to hold the printer.
- Bring extra supplies in case of accidents or failure. This includes spare filament, bed adhesive (e.g., painter tape), screwdrivers, and spatulas.
- You will need a laptop plus USB cable if the printer must print while tethered.
- If printing untethered, bring a spare SD card or USB drive with several models with varying print times.
- Bring an extension cord in case the power outlets are too far away for the power cable.

- Always relevel the printer's build plate after being moved.
- Before printing, check that all components and belts are secure, since pieces may loosen while being transported.

If you find yourself regularly taking the printer out for demonstrations, consider purchasing a small printer (see table 2.3 for suggestions) with a travel case for ease in portability.

Service Project Example

The University of Florida Marston Science Library is supporting the UF student club Generational Relief in Prosthetics (GRiP) to provide 3D printed prosthetic hands to local children with upper limb differences. This student organization is composed of undergraduate students who became involved in the Enabling the Future international project that matches volunteers with people who are missing part of their upper limbs. This project has several tested hand designs available for download freely online that can be tweaked and personalized to meet the needs of the recipient.

The library first became involved when librarians learned about Enabling the Future and printed out test hands, the first step in becoming a participant. At about the same time, GRiP students came to the library to print their own test hands and made the connection with the library. The librarians realized that this was a perfect opportunity to support a student group by assisting with modeling and printing guidance and to print the hands at no cost for the recipients. The 3D print lab at the science library is one of the largest on campus, so librarians continue to identify students who are interested in developing prosthetic hands and refer them to the GRiP team.

GRiP and the library are partnering in fall 2015 with the nearby Hands to Love camp, occupational therapists, and hand surgeons (http://www.handstolove.org/hand-camp) to match UF students with young campers. During visits to the camp and the library's 3D print lab, they will work together to customize a 3D printed hand (such as the Raptor Reloaded, http://www.thingiverse.com/thing:59696) for a child. The major objective is to teach the children and their families about 3D printing and assist them with modeling, printing, and assembling their own prosthetic hands.

Key Points

You will need to provide some level of training and marketing to build your 3D printing service. Tips include:

- Develop and present customized workshops with content appropriate for each group. Look for volunteer presenters.
- Point your users to existing videos and other forms of self-paced instruction.
- View invitations for road shows with caution, and be prepared when traveling with a 3D printer.
- Seek outreach and service opportunities that match your library's strengths with patrons' interests and strengths.

The next and final chapter will summarize the preparation steps for establishing a 3D printing service in your library. It will also highlight some exciting trends in 3D printing and remind you of ways to keep informed via the librarians' network.

◎ References

Buhler, Amy, Sara Russell Gonzalez, Denise Beaubien Bennett, and Erin R. Winick. 2015. "3D Printing for Middle School Outreach: A Collaboration between the Science Library and the Society of Women Engineers." Paper presented at the ASEE Annual Conference and Exposition, Seattle, WA, June 17. http://www.asee.org/public/conferences/56/papers/12660/view.

Fayetteville [NY] Free Library. 2014. "Makerspace FAQs: 21 Questions about FFL Makerspaces." Fayetteville Free Library. http://www.fflib.org/make/makerspace-faqs.

First Robotics Competition. 2013. "First Choice Adds Additives." *FRC Blog.* November 22. http://www.usfirst.org/roboticsprograms/frc/first-choice-adds-additives.

Harris, Jennifer, and Chris Cooper. 2015. E-mail message to the authors. July 9.

King, David Lee. 2015. "Landscape of Social Media for Libraries." *Library Technology Reports* 51 (1): 10–15.

Maker Faire. 2015. "Guidelines." Maker Media Inc. Accessed August 11. http://makerfaire.com/global/guidelines/.

McGoogan Library of Medicine. 2015. "3D Printing @ McGoogan Library: How to Print." McGoogan Library of Medicine. University of Nebraska Medical Center. Accessed August 11. http://unmc.libguides.com/content.php?pid=653308&sid=5410702.

Next Generation Science Standards. 2015. "Frequently Asked Questions." Achieve Inc. Accessed August 11. http://www.nextgenscience.org/frequently-asked-questions.

University of Texas, Arlington. 2015. "Outreach and K-12." UTA FabLab. University of Texas, Arlington. Accessed August 11. http://fablab.uta.edu/outreach.

Urban Libraries Council. 2015. "TechCentral MakerSpace." Urban Libraries Council. Accessed August 11. http://www.urbanlibraries.org/techcentral-makerspace-pages-330.php.

CHAPTER 12

Looking Ahead

IN THIS CHAPTER

▷ Bringing it all together

▷ Looking ahead

▷ Networking for guidance, advice, and support

DEVELOPING A 3D PRINTING SERVICE IS AN EXCITING CHALLENGE for libraries. 3D printing is a different kind of service from those traditionally offered, but it fits the mission of most libraries by fostering new knowledge, creativity, and collaboration. Space availability, budgetary issues, and staffing considerations make each situation unique, and this book has provided talking points, best practices, and examples from other libraries that will aid you in making wise decisions when planning a service for your library. Be sure to explore examples from all types of libraries, since service models, procedures, and policy points do not correlate by library type. Continue to examine other libraries' evolving documentation and join discussion lists to keep up with the industry and with evolving services offered in libraries.

Bringing It All Together

Before you launch the service, you and your staff must make decisions and train yourselves appropriately. The basic preparation tips described throughout this book are summarized as follows.

- Get staff buy-in before you launch a 3D printing service.
- Keep in mind that 3D printing itself is the end of a process that includes model selection or design.

- Prepare a space for the printer, its accessories, and related service areas.
- Obtain a printer, test it behind the scenes, and then begin your 3D service.
- Develop your policies.
- Decide whether to charge for printing.
- Train your staff.
- Choose your type of service: patron initiated, staff mediated, or both.
- Take advantage of knowledgeable patrons to assist with processing, printing, and workshop instruction.
- Brace yourself for print job failures.
- Use the librarians' network for advice and for keeping up with the industry.

Getting Staff Buy-In

To provide a successful service, your staff should accept this new and potentially disruptive service. Those who are not initially enthusiastic may profit from an awareness of how a 3D printing service fits with your mission and will benefit your patrons. At a minimum, all staff can learn to answer basic questions and to give appropriate referrals to colleagues who are more knowledgeable. Chapter 1 provides rationales used by other libraries, and chapter 9 offers tips on preparing as well as training your staff.

Obtaining a Model

Before you begin printing, you need a model file that describes a physical object in three dimensions. Patrons can find ready-made or customizable models from several repositories mentioned in chapter 3, or they can learn to make their own by using 3D scanners or modeling software as described in chapter 4. It is not critical for staff to develop modeling skills or to provide instruction in modeling, but any familiarity with the fundamental principles will result in better service provided to your patrons.

Preparing the Space

You must choose and prepare a space for the printer and its entourage, and the related expenses may be substantial. Also consider your service area needs and workspace for staff and for patrons. Since software is used to process the jobs, at least one computer must be dedicated or shared for this function, and placed at a location that is convenient for the type of service offered. Displays of sample prints of feature options, successes, failures, and filament choices will quickly take up space but are valuable for marketing purposes and as learning objects for patrons. Chapter 2 offers background to understand the space needs of a 3D printing service.

Testing the Printer

You don't need to know much about 3D printing before purchasing and unpacking a printer. Make as informed a decision as possible when selecting a printer, remembering that they are much like vehicles—they all work in the same basic manner, but the features and functionality will vary greatly, and personal preference plays a significant role in recommendations. Test your printer in a staff area until you achieve some confidence in its use, but don't wait for total mastery before you put it out for patrons. Chapter 2 on printers and chapter 8 on workflow will assist you in developing enough confidence with using your printer to move it into a visible location.

Choosing Your Type of Service

Choose whether you will offer patron-initiated, staff-mediated, a hybrid, or several types of service. Any of these options can work successfully in any type of library. Evaluate your service and workflow continually and don't be hesitant to change up the options at any time, as your staff and patrons gain experience and as you add more printers to your fleet. Chapter 2 describes how 3D printers work, and chapter 6 discusses some options for services.

Developing Your Policy

A 3D printing service policy will incorporate aspects of public policy issues as well as the local policies pertaining to similar library services. Rely on the American Library Association for guidance regarding public policy issues such as intellectual freedom, intellectual property, and liability. Many libraries have created local policies that govern their service, and examples provided throughout this book can be supplemented by checking websites and guides for updated statements. Coordinate this service policy with others in use at your library. Chapter 7 supplies examples of policies used at several libraries.

Charging for 3D Printing

The decision "to charge or not to charge" may be predetermined by your administration or by the size of your budget. If the choice is yours, first recognize that charge and noncharge payment models exist within each library type. Calculate whether you have the funding to continue to replace the supplies. Determine whether you can meet your patrons' demand for printing, or whether a control such as charging is appropriate. Chapter 5 guides you in estimating the expenses of running a 3D printing service and gives examples of charging models used by a variety of libraries.

Training Your Staff

After you have played with the printer, you will have a greater understanding of what tasks need to be assigned to staff. You may be surprised at which staff develop an early aptitude for 3D printing and are willing to train coworkers and to plan or assist with instruction. Chapter 8 provides examples of workflows, and chapter 9 describes staff preparation and training options so your staff can launch a service efficiently.

Relying on Your Patrons

3D printing is a skill in which patrons may know just as much, or even more, than library staff. You will find patrons who are happy to collaborate with your staff on understanding how your printer works. Some of your patrons may have significant expertise with modeling software, and they may be interested in leading workshops. Take advantage of these skill sets when planning your marketing and outreach activities as suggested in chapter 11.

Bracing for Disasters

Unfortunate things happen while 3D printing, as outlined in chapter 10. Poorly designed models, bad choices during job setup, misbehaving printers and filament, and other seemingly random problems will occur. Staff must be prepared to resolve problems, rerun jobs, and adjust queues and reservations as needed.

⊚ Looking Ahead

It can be challenging to forecast the direction 3D printing will take in the future, due to unpredictable factors such as new advancements in technology and unexpected funding sources (e.g., Kickstarter) that encourage nontraditional manufacturers. However, in the near future, the following 3D printing topics will be a focus of development and should be considered when planning for your service.

Ubiquity of 3D Printers

3D printers are now penetrating the consumer market, with models being sold in high-traffic brick-and-mortar stores such as Home Depot and Sam's Club (McCue, 2015). As the technology matures, components drop in price and the software becomes more user friendly. The goal is for 3D printers to become another piece of equipment, like 2D printers or scanners, which average consumers now purchase for home use. Challenges for home use include air quality and maintenance issues as addressed throughout this book.

Although patrons may eventually have such technology at home, librarians know from previous experience that patrons still come to libraries or to commercial businesses to use other items, such as photocopiers, to use larger or fancier versions of equipment than those found at home, or to use equipment not available in the home. 3D printers may reach the home market for hobbyists or for some professionals, such as engineers or artists. But the library may continue to serve as a place where people are introduced to 3D printers, where patrons can use a 3D printer whose features surpass their own, or where patrons can obtain access when nothing is available in the home. If 3D printers do become commonplace, this market should spur development of new technology, materials, software, and support networks.

Already, networks of 3D printers are forming, in which local home- or small-business-based makers print models on demand. Examples of these distributed networks are 3DHubs.com and PrintAThing.com. These services benefit both the makers who wish to monetize their own 3D printers and consumers with have a limited need for 3D printing and do not want to purchase and maintain a 3D printer. Libraries should consider participating in similar networks, even without the goal of increasing revenue, to notify potential users of available 3D services.

As 3D printers become more commonplace, local and regional support services will eventually form to repair and supply 3D manufacturers. Currently, other than a few physical retailers that are located mainly in large cities, all equipment and supplies must be purchased online. This includes the consumables such as filament, and repair options, which are currently either do-it-yourself or mailing off the machine or specific components to the company.

As the 3D printing industry matures, look for opportunities for entrepreneurs to supply and support local 3D printers. Large- and small-scale manufacturers can make and sell filament and other starting materials. The current hurdles in recycling filament scraps into reusable filament, as discussed in chapter 2, are likely to be resolved in the long term. A 3D printer repair industry should sprout, similar to that developed for repairing personal computers and cell phones. Owners will tire of returning a printer to the manufacturer at each point of failure. Those on your staff who excel with technical support of your 3D printers may find several opportunities for moonlighting opportunities.

3D Material

As discussed in chapter 2, 3D material manufacturers are continually exploring new filament materials. The company Taulman 3D, for example, is focused on high-performance 3D printing and is developing filaments designed to withstand high stress and temperatures (Taulman 3D, 2015). These filaments are especially appealing to engineers who need stronger materials that shrink minimally during printing. Eastman Chemical Company recently introduced Amphora, a new polymer for 3D printing, that promises strength, gloss, and low-odor emission that should make it attractive for users with 3D printers in public areas (Eastman Chemical Company, 2015). While new materials will be introduced that will meet the specific needs of some patrons, libraries must remain cognizant that these materials may not be suitable for their printers or environmentally safe for public areas.

Manufacturers are also creating composites of plastic plus additives such as wood and metal particles. ColorFabb currently sells bamboo, bronze, brass, and copper (ColorFabb, 2015) and other companies, such as 3Dom USA and MakerBot, follow this filament trend with their own creative composite product lines (Dotson, 2015; CNET.com, 2015). Expect this trend to continue with companies incorporating unique particulates within existing types of plastics such as PLA.

While color shades aren't technologically difficult to produce, consumers will see custom lines of colors, such as MakerBot's Martha Stewart line (Steele, 2014). Glow-in-the-dark filament is appealing to patrons, but it is costlier than standard filament. Testers reveal that the pricier glow filaments have more "glow" pigments, and the standard green color has better glowing properties than the newer additional colors (3dprintingforbeginners.com, 2015). In response to the desire to have color variations without dual extruders, manufacturers are developing filament that changes color when exposed to light or temperature or has color gradients across the length of the spool (Mendoza, 2015).

For users that want to print with multiple colors or materials, the current solution is to use a multiextruder 3D printer as discussed in chapter 2. Inventors are developing devices for single-extruder printing with multiple filaments, such as the Palette, a device that switches up to four types of filament in a single model (Mills, 2015). Another option is the Diamond Hotend, an extruder that allows users to print with at least three different colors of filament (RepRap.org, 2015). These filament switchers require modifications to slicing software and are not compatible with all 3D printers; however, these and future devices are promising for users who do not wish to manage multiextruder printers.

Software Developments

Modeling Software

Expect to see modeling software programs that are especially designed for specific purposes, such as audiences or functions. Examples include easy-to-use software for new or occasional users, or software specifically geared for 3D printing usage. Autodesk is an example of a company developing an ecosystem of software targeting various audiences. Their new product, Tinkerplay, is a mobile app that interfaces with Tinkercad.com to allow users to design and print 3D characters (Paul, 2015). Tinkerplay illustrates a trend toward expanding modeling software from desktop versions to web-based and mobile apps, including tools that serve a professional audience. Examples of these powerful cloud-based CAD tools include Onshape and Fusion 360 (Maxey, 2015). The proliferation of online modeling tools, especially those with free user accounts, will be beneficial to libraries and schools who do not have the resources to purchase expensive software licenses.

Queue Management Software

As usage increases and equipment prices decrease, libraries may find that transaction costs are limiting their service and seek alternatives to minimize the labor needed to manage transactions. John Hauer of 3DLT predicts that "2015 will be the year of workflow. Software companies, equipment suppliers, and service bureau customers will develop, sell, and buy software that makes 3D printing more productive and efficient at scale" (Hauer, 2014). Look for libraries to build and share their locally developed queue management software, as described in chapter 8, or adopt commercial products that manage workflow and facilitate patron transactions.

Diversity in the Marketplace

The 3D printing industry may branch into several routes to serve a variety of users. 3D printers for specialized industries, such as food production and medicine, will continue to develop and expand into niche sectors. Hobbyists and boundary pushers will seek out 3D printers with new features and capability, prioritizing function over tested reliability. Production service operations, including libraries, will focus on the options of reliable and indestructible workhorse printers that steadfastly print out models for patrons without endless repairs or calibration.

In the library world, a tension will exist among the patrons who want more advanced 3D printers with a wide variety of print materials, the patrons with basic needs who just require ready access to a functional 3D printer, and the staff who acknowledge the challenges of providing limited services in a public space. Whatever happens to the industry, libraries are likely to pursue the dependable models that churn out a modest but reliable quality of print jobs. At some point in the far future, 3D printers may become as common as 2D printers, and libraries may continue to provide them for patrons lacking access or for convenience, as they already do with other office-type equipment.

Using the Librarian's Network

Finally, keep in mind that you are not the only library running a 3D printing service! Colleagues in all types of libraries are ready to provide opinions about technology and service. Sign up for discussion lists, read their archives, and ask your questions. As a reminder, two helpful lists are:

Librarymakerspace-L@lists.ufl.edu

To join, send an e-mail message to listserv@lists.ufl.edu.

In the body of the message, type: subscribe librarymakerspace-l <your name>

LITA (American Library Association / Library and Information Technology Association) 3D Printing Interest Group Mailing List

To join, go to http://lists.ala.org/sympa/info/lita-3d and follow the instructions.

Continue to peruse the guides and websites of other libraries. The list of libraries with policies presented in chapter 7 is a good place to start. Examples presented in this book will become outdated as libraries acquire additional equipment, grow their service, or refine their policies.

Key Points

This book has provided examples of several aspects of running a 3D printing service that are followed in many types of libraries. If you are planning to implement your own 3D printing service, please note the following points of encouragement:

- Any library can support a successful 3D printing service.
- Preparation can pave the way for acceptance.
- A 3D printing service fosters discovery and creativity, fitting well with most libraries' missions.
- The 3D printing industry and library 3D printing services will evolve rapidly. Make sure you are following up-to-date information.

Any library can develop and launch a successful 3D printing service, bringing a sense of excitement and new patrons into the library. A 3D printing service provides an opportunity for your patrons to develop practical solutions for many of their needs, and it reinforces the library's role as a place of discovery.

References

3dprintingforbeginners.com. 2015. "What Material Should I Use for 3D Printing? Advanced Materials Review #4—Glow in the Dark Filament." *Materials* (blog). August 25. http://3d printingforbeginners.com/glow-in-the-dark-filament-review/.

CNET.com. 2015. "MakerBot Makes 3D Printing More Realistic with Metal, Wood and Stone." *CES 2015* (blog). January 6. http://www.cnet.com/news/makerbot-to-add-composite-mate rials-professional-service-and-remote-monitoring-to-its-5th-gen-3d/.

ColorFabb. 2015. "Specials." ColorFabb. Accessed September 20. http://colorfabb.com/specials.

Dotson, Kyt. 2015. "Coffee-Based Filament Allows People to 3D Print Beautiful, Brown Objects." *SiliconANGLE* (blog). SiliconANGLE Media. August 24. http://siliconangle. com/blog/2015/08/24/coffee-based-filament-allows-people-to-3d-print-beautiful-brown-ob jects/.

Eastman Chemical Company. 2015. "Better Air Quality for 3D Printers Is a Beautiful Thing." *Marketing Publication* SP-MBS-1113. July. http://www.eastman.com/Literature_Center/S/ SPMBS1113.pdf.

Hauer, John. 2014. "A Big Prediction for 3D Printing in 2015." *3DLT Blog*. 3DLT LLC. December 18. http://blog.3dlt.com/2014/12/18/a-big-prediction-for-3d-printing-in-2015/.

Maxey, Kyle. 2015. "Cloud-Based CAD Getting Traction." *Designer Edge* (blog). ENGINEER-ING.com. July 24. http://www.engineering.com/DesignerEdge/DesignerEdgeArticles/ ArticleID/10464/Cloud-Based-CAD-Getting-Traction.aspx.

McCue, T. J. 2015. "3D Printers Popping Up in Retail Stores." *Tech* (blog). Forbes.com LLC. April 17. http://www.forbes.com/sites/tjmccue/2015/04/17/3d-printers-popping-up-in-retail-stores/.

Mendoza, Hannah Rose. 2015. "Stronghero Introduces PLA Filament for 3-Color Gradient 3D Printing." *3D Printing Materials* (blog). 3Dprint.com. May 15. http://3dprint.com/65288/stronghero-gradient-filament/.

Mills, Chris. 2015. "This Box Turns Any 3D Printer into a Multi-Color, Multi-Material Marvel." *3-D Printing* (blog). Gizmodo. April 21. http://gizmodo.com/this-box-turns-any-3-d-printer-into-a-multi-color-mult-1699332022.

Paul, Ian. 2015. "Autodesk's Insanely Fun Tinkerplay App Lets You Easily Design Your Own 3D-Printable Action Figures." *Graphics and Design* (blog). PCWorld. March 18. http://www.pcworld.com/article/2898185/autodesks-insanely-fun-tinkerplay-app-lets-you-easily-design-your-own-3d-printable-action-figures.html.

RepRap.org. 2015. "Diamond Hotend." *RepRap* (wiki). September 10. http://reprap.org/wiki/Diamond_Hotend.

Steele, Billy. 2014. "MakerBot, Martha Stewart Decorate Parties with 3D-Printed Goods." *Engadget* (blog). AOL Inc. November 17. http://engadget.com/2014/11/17/makerbot-martha-stewart-3d-printing/.

Taulman 3D. 2015. "High Strength Materials for Your 3D Printer." Taulman3D LLC. http://taulman3d.com.

Glossary

3D model—A digital representation of an object created by **3D modeling software.** Models designed for 3D printing often are just the surface (or **shell**) of an object.

3D model libraries (or repositories)—Sites where creators can upload their 3D models for users to download and print themselves. Popular sites include Thingiverse, GrabCAD, and YouMagine.

3D modeling software—Computer graphics software used to create a 3D model that may be exportable for 3D printing. Common programs include Tinkercad (great for beginners), Google SketchUp, CAD (computer-aided design) programs such as Autodesk Inventor and SolidWorks (used by engineers), Blender, Maya, and Rhino (used by artists). Some programs are free to use upon registration.

3D printer—The machine that follows **slicing** or **G-code** instructions to create an object from a **3D model.** Less commonly used for **additive manufacturing** machines that process metalworking.

3D scanner—A device that analyzes and collects data about an object that can be used to construct a **3D model.** Scanners are available in stand-alone or handheld versions.

ABS (acrylonitrile butadiene styrene)—Petroleum-based type of plastic filament commonly used in the **FFF** 3D printing process.

additive manufacturing—The umbrella term for prototyping processes whereby a real object is created from a 3D design. The digital **3D model** is saved in **STL** format and then sent to a 3D printer. The 3D printer then prints the design layer by layer to form a real object.

brim—a ridge added to the base of the model for several layers to add stability while printing. The brim, like the **raft**, is designed to break off from the model after removal from the build plate.

build plate—The flat surface on which the material is deposited to build a 3D model. Build plates are commonly made out of glass or acrylic. Also called the print bed.

build size—The maximum dimensions (length, width, height) that a 3D printer can print. 3D models that are larger than these dimensions cannot be printed. Also known as the build envelope.

dual extrusion—Refers to a 3D printer with two extruder heads to provide the capability of printing with multiple colors or types of filament at the same time. There are 3D printers with three and four extruder heads as well.

extruder—The part of the 3D printer that heats and deposits a thin strand of **filament.**

filament—The wirelike material used in **FFF**-style 3D printers. Most filament is plastic based, available in multiple colors, and 1.75 or 3.0 mm size in diameter. Typically wound into spools although some printers require filament housed in cartridges.

FFF (fused filament fabrication)—A form of additive manufacturing in which a thin **filament** is heated and pushed through a **nozzle**, depositing thin layers onto a flat build plate. This process continues as a model is built up from the **build plate** layer by layer. The most well-known type of 3D printing because it dominates the hobby and "prosumer" markets. FDM (Fused Deposition Modeling) is synonymous but is a proprietary term owned by Stratasys Ltd.

G-code—A text file produced by the **slicing** software that contains instructions for producing each slice or layer of the item, including where and how fast to move and what path to follow. Files generated by proprietary 3D printer software have highly variable file extension names, such as X3G and S3G.

infill—The interior pattern and density of a 3D model that is controlled by the printer software. The density of the infill pattern (e.g., honeycomb, rectilinear) is typically 10–30 percent, but both pattern and density can be adjusted to provide greater model strength and weight.

layer height—The thickness of an individual layer in a 3D model (e.g., 0.20 mm). Considered as an indication of **resolution** with thinner layer heights corresponding to higher resolution.

makerspace—A community-operated workspace that typically contains machines and tools as well as knowledge that can be shared by users. Some 3D printing labs are called makerspaces even if they don't contain any other types of tools or equipment.

mesh—The exterior surface of a 3D digital model. **STL** files describe the mesh using triangles connected at their vertices, with smaller triangles providing higher resolution. Missing triangles indicate a hole or error in the mesh that will need to be corrected using **model repair software**.

model repair software—Used to verify that a 3D model is geometrically sound (free of **mesh** or manifold errors) and therefore printable. Files that have been produced through **3D scanning** tend to have more errors than files created through **modeling software**. Examples include Microsoft Model Repair Service, MeshLab, Meshmixer, MiniMagics, and Netfabb. Some basic-level services are free to use upon registration.

nozzle—The exit point of the **extruder** for the heated material.

OBJ—Abbreviation for Object file (OBJ or .obj). A common 3D model file format.

overhang—Part of a model that protrudes at an angle ≥ 45 degrees as measured from the vertical. Overhangs will typically require added **support** to successfully print.

PLA (polylactic acid)—Plant-based plastic used in **FFF**-style 3D printing. Biodegradable, extrudes at a lower temperature than **ABS**, and doesn't require a heated build plate.

raft—A thin layer of filament, slightly larger than the model's footprint, deposited before printing the actual object. Raft supports are constructed to be removable, either by pulling off or by dipping in a chemical bath.

rapid prototyping—Iterative cycle that quickly allows an inventor to go from idea to a physical object that is tested and remade with improvements before final or mass production. 3D printing can be an essential component of this process since it allows a single model to be built more rapidly than possible using other manufacturing processes.

resin—Liquid photopolymer that is formulated to harden when exposed to UV or visible light. Used in SLA-type 3D printers such as Formlabs' Form 1.

resolution—The minimum feature size that can be expected to be reproduced, as determined by the thickness of the filament extruded. For each printer, "high" or "low" or "standard" resolution correspond to the **layer height**.

scanner/scanning—Refers to a 3D scanner. 3D objects must be scanned with a 3D scanner to generate a likeness that can be reproduced in three dimensions.

shell—The outer layer or skin of a 3D printed model. Increasing the number of shells adds rigidity and strength to a model.

skirt—A thin outline of the model's footprint, deposited on the plate before beginning the actual printing.

slice—A thin cross section of the 3D printed model. Software to prepare the model for printing "slices" the model into many layers and determines the path of the extruder to build each layer.

SLA (stereolithography)—An additive method of light polymerized 3D printing that hardens liquid resin with near-UV lasers.

SLS (selective laser sintering)—An additive method of granular 3D printing that fuses fine powders, layer by layer, into 3D shapes.

STL—Short for stereolithography or standard tessellation language. A file format from 3D modeling programs commonly used in 3D printing. Written as both "STL" and ".stl."

supports—Temporary structure added to a 3D models to buttress **overhangs** or bridges.

x, y, z axes—Three-dimensional coordinate system. The x axis represents the "horizontal" or length of the item, y is the depth, and z is the "vertical" or the height of the item.

Index

Note: Page references for figures are italicized. Please also consult the glossary.

University, 5; Fayetteville [NY] Free Library, 7, 72, 76, 89, 106, 118, 121, 126, 129, 148; Gloucester County (NJ) Library System, 89, 121, *149*; Hinsdale (IL) Public Library, 96, 100; Illinois Institute of Technology, 76, 89, 96–97, 100, 110; Innisfil [Ontario] Public Library, 2; Keene (NH) Public Library, 3, *149*; Kent State University, 6, 89; Lake Forest Academy (IL), 86–87, 89, 96; Lewes (DE) Public Library, 96; Makerspaces in Idaho Libraries Project, 3; McGoogan Library (University of Nebraska), 88–89, 95, 143, 148–49; Miami University (OH), 91; North Carolina State University, 73, 91, 110; Northbrook (IL) Public Library, 49, *149*; Northeastern University, 75; Oswego State University of New York, 91; Riverdale Collegiate Institute (Toronto, ON), 86, 91, 103–4, *149*; Sacramento (CA) Public Library, 3, 96; San Diego Public Library, 7, 71, 76, 85, 91, 94; Southern Illinois University Edwardsville, 91; Southern New Hampshire University, 3, 4, 46, 151–52; Stetson University, 6; St. Joseph County (IN) Public Library, *149*; Tulsa City-County Library, 103; University of Alabama, 76, 87–88, 91, 121, 122–23; University of Arizona, 3, 49, 110; University of Florida, 5, 7, 70–71, 75, 76–77, 87, 91, 99–100, 110, 122, 127, 129, 134, 135, 143–145, 152, 154; University of La Verne, 4, *5*; University of Maryland, 88, 91, 110; University of Memphis (TN), 91, 105; University of Michigan, 73, 91, 105, 106; University of Nebraska Medical Center. *See* McGoogan Library; University of Nevada Reno, 91; University of North Texas, 76, 91; University of Tennessee-Knoxville, 91; University of Texas at Arlington, 153; University of Utah, 49, 91, 104–5; University of Wisconsin–Stevens Point, 91, 95, 97, 104; Urban Libraries Council. *See* Cleveland Public Library; Warwick [RI Public] Library, 76; Westport [CT Public] Library, 72, 85–86, 91, 104
LIBRARYMAKERSPACE-L, *9*, 28, 162
location of printer, 18–20

maintenance and repair, 129–30. *See also* troubleshooting
marketing, 19, 66, 118, 147–49
model: flawed, 43–44, 70, 75, 95, 123–24, 139; design principles, 44–49; design software, 37–41, 75, 124, 161–62; filetypes, *33*; repair software, 50, *51*, 95; repositories, 32–36, 95; to be printed, 158
multiple printers, 106–7
multiple service models, 73

nearby 3D printing services, 6, 23–24
noise, 20
notification of completion, 102, 125–26

OBJ (.obj), 32, 41
options to library 3D printing. *See* alternative printing services
outreach, 153–54; examples, 153, 154. *See also* marketing

patents, 12
patron expectations, 140–41
patron-managed service, 71–72; academic example, 72, 143; public library example, 72; troubleshooting, 140; workflow, 102–5
PLA (polylactic acid), 15, 20
placement of printer. *See* location of printer
policies, 81, 159; aligned with others, 82–83; examples, 84–91; unique, 83–84; updating, 91. *See also* public policy
primer reading list, *13*
print job setup, 50–55, 98, 122–23, 136–38. *See also* file preparation and processing
printer costs, 61, *62*
printer features, 22–25, 162
printer selection, 21–28
printer size, 21
printer and processing software, 24, 107, 110–11, *112*, 162
prioritizing print jobs, 78–79, 83, 114, 125
problem patrons, 141–42
prosthetic hands, 154
public policy, 81–82. *See also* intellectual freedom and property, liability
publicity. *See* marketing

queue management, 78–79, 100–102, 106, 125, 162

rafts, 52–54, 101
rationale, 1–2, 118; examples, 2–3
recycling (filament), 16–17, 19, 101
remote monitoring, 19
replacement parts, 23, 65–66
repositories. *See* model, repositories
review sources, 28

safety, 20, 76, 84, 85–86, 87, 90
scale, 39, 46, 98
scanning. *See* 3D scanners

About the Authors

Sara Russell Gonzalez is the physical sciences, mathematics, and visualization librarian at the Marston Science Library at the University of Florida. She holds a BS from Caltech in geophysics and a PhD from the University of California, Santa Cruz, in seismology, and an MLIS from Florida State University. A former geophysicist, her research interests include emerging technologies in libraries, modeling and visualization of data, and scientific literacy instruction. She coordinates the Marston Science Library Visualization Room, the MADE@UF software development lab, and the 3D printing service with Denise Bennett.

Denise Beaubien Bennett is a librarian emeritus after serving for over thirty years as an engineering librarian at the Marston Science Library at the University of Florida. She holds AB and AMLS degrees from the University of Michigan. She has held many offices in divisions of ALA, including RUSA, the ACRL Science and Technology Section, and MARBI. She served as general editor for ALA's *Guide to Reference* from 2009 to 2015, and as the division editor of its Science, Technology, and Medicine Section from 2005 to 2015. Her professional awards include the RUSA MARS Certificate of Appreciation ("My Favorite Martian") in 2002, the Reference Service Press Award in 2006, and the RUSA Isadore Gilbert Mudge Award in 2015. Her research interests include database searching, indexing, and evaluation; science librarianship; evaluation of reference tools; and responsible conduct of research (RCR) training. Her latest service interest is implementing 3D printing in academic libraries.

Since obtaining a 3D printer in 2014, Sara and Denise have collaborated on training activities for patrons and for librarians. Their joint publications and presentations on 3D printing in libraries are listed at http://guides.uflib.ufl.edu/library3Dprinting/aboutus.